Math Mammoth
Grade 6 Answer Keys

for the complete curriculum
(Light Blue Series)

Includes answer keys to:

- Worktext part A
- Worktext part B
- Tests
- Cumulative Reviews

By Maria Miller

ISBN 978-1-942715-23-8

EDITION 12/2016

Contents

Math Mammoth Grade 6-A

Answer Key

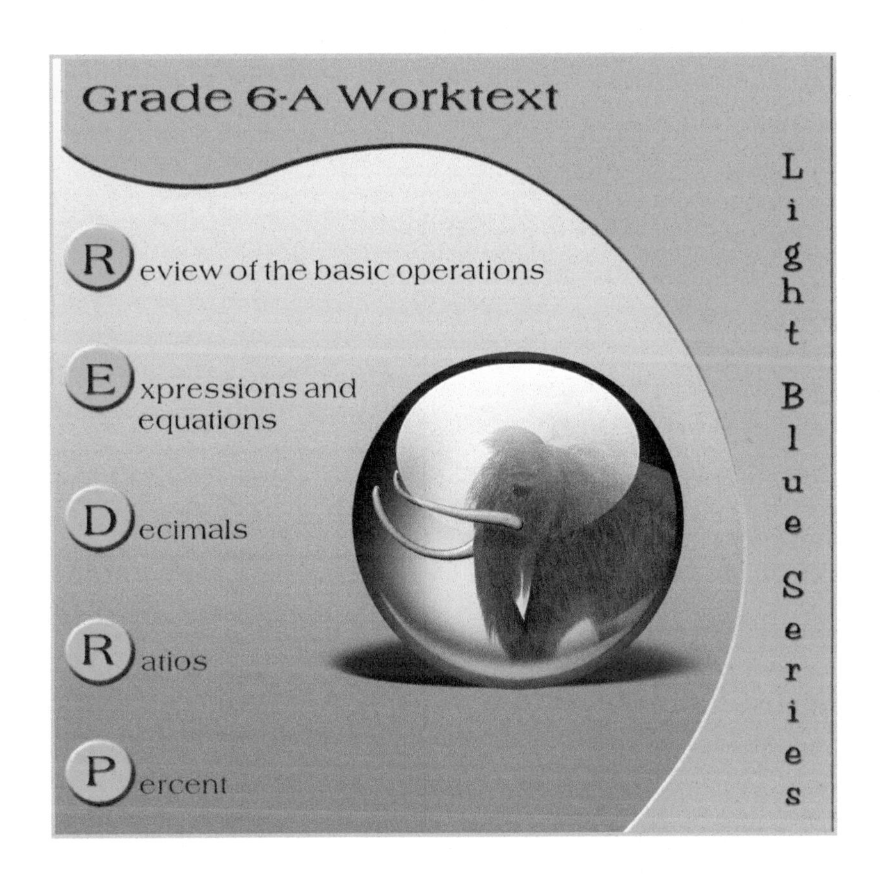

By Maria Miller

EDITION 12/2016

Math Mammoth Grade 6-A Answer Key

Contents

Chapter 5: Percent

Chapter 1: Review of the Basic Operations

Warm-up: Mental Math, p. 9

1. a. 410 + 38 = 448
   b. 150 + 600 = 750
   c. 5,600 − 2,000 = 3,600
   d. 14 + 90 = 104
   e. First, add 45 and 56 to get 101. Then, 101 + 35 = 136.
   f. 60 − 40 = 20

2. a. 93 b. 655 c. 380

3. a. 60 c. 8 e. 560 g. 21 i. 6
   b. 72 d. 50 f. 40 h. 200 j. 9

4. a. 500 − 40 − 150 = 310
   b. 1,020 − 30 × 20 = 1,020 − 600 = 420
   c. 30,000 + 15,000 = 45,000
   d. 50 × 70 = 3,500
   e. 105 + 45 = 150
   f. 1,000 + 90 = 1,090

5. a. x = 2,800
   b. M = 60
   c. y = 180

6.

n	130	250	360	410	775	820	1,000
n − 35	95	215	325	375	740	785	965

7.

n	3	5	12	15	25	35	60
n × 40	120	200	480	600	1,000	1,400	2,400

8. a. Each piece of board is 110 cm long: (600 cm − 50 cm) ÷ 5 = 550 cm ÷ 5 = 110 cm.

9. a. Eve's hourly wage is $104.00 ÷ 8 = $13.00 per hour.
   b. Eve earns $104 × 5 = $520 in a week, and $520 × 13 = $6,760 in three months.

10. a.
    6 3/4 cups of flour
    1 cup of honey
    1 1/2 cups of butter
    2 1/4 teaspoons of nutmeg
    4 1/2 teaspoons of cinnamon
    1 1/2 teaspoons of ground cloves
    2 1/4 cups of walnuts

    b. They made 7 1/2 dozen or 90 cookies.

Review of the Four Operations 1, p. 11

1\. a.

```
   2 4
   5 3 6
 x   7 1
   5 3 6
 3 7 5 2
 3 8,0 5 6
```

b.

```
  $2 4.5 9
 x    7 0
$1,7 2 1.3 0
```

c.

```
      2 0 6
    x 9 1 5
    1 0 3 0
    2 0 6
1 8 5 4
1 8 8,4 9 0
```

d.

```
   1 3 8 7
4)5 5 4 8
  4
  1 5
  1 2
    3 4
    3 2
      2 8
      2 8
        0
```

e.

```
   8 5.6
7)5 9 9.2
  5 6
    3 9
    3 5
      4 2
      4 2
        0
```

f.

```
   1 0 3 4
8)8 2 7 2
  8
  0 2 7
    2 4
      3 2
      3 2
        0
```

2\. Multiply the quotient by the divisor to check your division.

```
  1 3 2           3 4           2 3
  1 3 8 7         8 5.6       1 0 3 4
x       4     x       7     x       8
  5 5 4 8       5 9 9.2       8 2 7 2
```

3\. a. 6,048
b. 34.95
c. 109,841 R2

4\. a. 37 ÷ 6 = 6 R1; 6 × 6 + 1 = 37
b. 54 ÷ 5 =10 R4; 5 × 10 + 4 = 54
c. 61 ÷ 8 = 7 R5; 8 × 7 + 5 = 61

5\.

2 × 45 = 90 3 × 45 = 135 4 × 45 = 180 5 × 45 = 225 6 × 45 = 270 7 × 45 = 315 8 × 45 = 360 9 × 45 = 405	a. 45) 4 0 0 5 = 8 9 -3 6 0 4 5 - 4 5 0	3 4 8 9 × 4 5 4 4 5 + 3 5 6 0 4 0 0 5
2 × 30 = 60 3 × 30 = 60 4 × 30 = 90 5 × 30 = 120 6 × 30 = 180 7 × 30 = 210 8 × 30 = 240 9 × 30 = 270	b. 30) 2 0 2 2 0 = 6 7 4 -1 8 0 2 2 2 - 2 1 0 1 2 0 - 1 2 0 0	2 1 6 7 4 × 3 0 2 0 2 2 0
2 × 75 = 150 3 × 75 = 225 4 × 75 = 300 5 × 75 = 375	c. 75) 1 9.8 7 5 = 0.2 6 5 -1 5 0 4 8 7 - 4 5 0 3 7 5 - 3 7 5 0	3 4 0.2 6 5 × 7 5 1 3 2 5 + 1 8 5 5 0 1 9.8 7 5

Review of the Four Operations 1, cont.

6. a. 1,813 R1; 48 × 1813 + 1 = 87,025
   b. 9,685 R10; 90 × 9685 + 10 = 871,660
   c. 658 R66; 82 × 685 + 66 = 54,022

7.

a. 2,960 R86 2960 101) 299046 -202 970 -909 614 -606 86	34 2960 × 101 2960 0 +296000 298960 + 86 299046
b. 29,546 R48 29546 123) 3634206 -246 1174 1107 672 -615 570 -492 786 -738 48	29546 × 123 88638 590920 +2954600 3634158 + 48 3634206
c. 21,862 R300 21862 350) 7652000 -700 652 -350 3020 -2800 2200 -2100 1000 -700 300	21862 × 350 0 1093100 +6558600 7651700 + 300 7652000

Review of the Four Operations 1, cont.

8. **I** 42,408 ÷ 76 = 558
M 85,104 ÷ 54 = 1576
O 23,530 ÷ 26 = 905
R 61,880 ÷ 35 = 1768
V 51,944 ÷ 86 = 604
E 44,217 ÷ 51 = 867
I 223,496 ÷ 91 = 2456
I 30,624 ÷ 33 = 928
R 133,140 ÷ 70 = 1902
S 11,880 ÷ 22 = 540
E 128,316 ÷ 111 = 1156
E 51,313 ÷ 97 = 529
M 880,341 ÷ 309 = 2849
T 113,168 ÷ 88 = 1286
R 693,360 ÷ 810 = 856

What is as round as a dishpan, and no matter the size, all the water in the ocean cannot fill it up? SIEVE

What flies without wings? TIME

I am the only thing that always tells the truth. I show off everything that I see. MIRROR

G 200,196 ÷ 201 = 996
O 324,729 ÷ 57 = 5697
E 339,388 ÷ 31 = 10948
S 2,337,820 ÷ 205 = 11404
E 28,548 ÷ 18 = 1586
R 617,105 ÷ 415 = 1487
S 2,863,250 ÷ 250 = 11453
T 1,049,664 ÷ 88 = 11928
H 236,215 ÷ 35 = 6749
F 97,920 ÷ 16 = 6120
O 1,388,740 ÷ 230 = 6038
P 759,290 ÷ 70 = 10847
I 678,040 ÷ 506 = 1340
T 250,536 ÷ 44 = 5694
F 239,397 ÷ 199 = 1203

From what heavy seven-letter word can you take away two letters and have eight left? FREIGHT

The more of them you take, the more you leave behind. What are they? FOOTSTEPS

Review of the Four Operations 2, p. 17

1. a. $29,600 + $13,500 + $8,300 = $51,400. $51,400 ÷ 4 = $12,850.
The family used $12,850 for groceries.
b. 1/5 + 1/4 = 4/20 + 5/20 = 9/20. The family had 11/20 of their income left after taxes and groceries.

2. a. 100 − 29.5 × 2.6 = 100 − 76.7 = 23.3
b. 2.3 + 9.356 + 0.403 + 908.8 = 920.859
c. 800 − (12.48 − 2.9) = 800 − 9.58 = 790.42
d. 559.50 ÷ 3 = 186.5

3. 4,958 ÷ 13 = 381 R5 OR 4,958 ÷ 381 = 13 R5

4. a. You would need to add *four* zeroes so that you can calculate the dividend to four decimal digits. You will need four decimal digits in order to round it to three decimal digits.
b. 65.0000 ÷ 7 = 9.2857, which rounds to 9.286.

5. 10 m × 12 m = 120 m^2 ; 120 m^2 ÷ 9 = 13.33 m^2. The area of each section is 13.33 m^2.

6. The farmer needed 262 boxes to pack the apples. Notice the problem doesn't tell you *how many apples* there were, but instead tells you how many kilograms of apples there were. Since four apples make a kilogram, he had 2,350 × 4 = 9,400 apples. Now divide: 9,400 ÷ 36 = 261 R4. He needed 262 boxes.

7. a.

Miles	9	18	27	54 miles	108	135	162
Time	10 min	20 min	30 min	1 hour	2 hours	2 1/2 hours	3 hours

b. They will travel 486 miles.

c. It will take them approximately ten hours to travel 550 miles.

8. a. At 40 mph, it takes him 1.5 minutes to drive each mile. You can solve this in many ways. For example, since he drives 40 miles in 60 minutes, you can make a table like in exercise 7, and find that he drives 20 miles in 30 minutes, 10 miles in 15 minutes, and 5 miles in 7 1/2 minutes.
b. Twenty times as long as what it takes him to drive 5 miles: 20 × 7.5 minutes = 150 minutes.
c. To drive 30 miles will take him six times as long as to drive 5 miles, so it takes him 6 × 7.5 minutes = 45 minutes. Dad would have to leave at 8:15 a.m. to arrive at 9 a.m.

9. a. One gallon is 128 liquid ounces. Ninety-six gallons is 96 × 128 oz = 12,288 oz. and 12,288 ÷ 8 = 1,536.
They filled 1,536 eight-ounce bottles with fruit juice.
b. $3,072 ÷ 1,536 = $2. They would have to charge at least $2 per bottle to break even.

Puzzle corner: a. 4,392 − 293 + 293 = 4,392 b. 384 ÷ 8 × 8 = 384 c. $\frac{1,568}{49} \times 49 = 1,568$

Powers and Exponents, p. 20

1. a. $3^2 = 3 \times 3 = 9$
   b. $1^6 = 1 \times 1 \times 1 \times 1 \times 1 \times 1 = 1$
   c. $4^3 = 4 \times 4 \times 4 = 64$
   d. $10^4 = 10 \times 10 \times 10 \times 10 = 10{,}000$
   e. $5^3 = 5 \times 5 \times 5 = 125$
   f. $10^2 = 10 \times 10 = 100$
   g. $2^3 = 2 \times 2 \times 2 = 8$
   h. $8^2 = 8 \times 8 = 64$
   i. $0^5 = 0 \times 0 \times 0 \times 0 \times 0 = 0$
   j. $10^5 = 10 \times 10 \times 10 \times 10 \times 10 = 100{,}000$
   k. $50^2 = 50 \times 50 = 2{,}500$
   l. $100^3 = 100 \times 100 \times 100 = 1{,}000{,}000$

2. a. $2^6 = 64$
   b. $8^5 = 32{,}768$
   c. $40^2 = 1{,}600$
   d. $10^4 = 10{,}000$
   e. $9^8 = 43{,}046{,}721$
   f. $11^3 = 1{,}331$

3. a. 144 km^2
   b. A = 6 m × 6 m = 36 m^2
   c. A = 6 in × 6 in = 36 in^2
   d. A = 12 ft × 12 ft = 144 ft^2

4. a. 2 cm × 2 cm × 2 cm = 8 cm^3
   b. V = 10 in × 10 in × 10 in = 1,000 in^3
   c. V = 1 ft × 1 ft × 1 ft = 1 ft^3
   d. V = 5 m × 5 m × 5 m = 125 m^3

5. a. If the perimeter (four sides) is 40 cm, then one side is 10 cm. So the area is 10 cm × 10 cm = 100 cm^2.
   b. Since $4^3 = 4 \times 4 \times 4 = 64$, a cube with a volume of 64 in^3 has sides that are 4 in long.
   c. If the square's area is 121 m^2, then the length of one side is 11 m. So the perimeter is 4 × 11 m = 44 m.
   d. Since $3^3 = 3 \times 3 \times 3 = 27$, a cube with a volume of 27 cm^3 has sides that are 3 cm long.

6.

a.	b.	c.	d.
$2^1 = 2$	$3^1 = 3$	$5^1 = 5$	Answers will vary. Please check the students' work.
$2^2 = 4$	$3^2 = 9$	$5^2 = 25$	
$2^3 = 8$	$3^3 = 27$	$5^3 = 125$	
$2^4 = 16$	$3^4 = 81$	$5^4 = 625$	
$2^5 = 32$	$3^5 = 243$	$5^5 = 3{,}125$	
$2^6 = 64$	$3^6 = 729$	$5^6 = 15{,}625$	

7. a. Since $3^7 \times 3 = 3^8$, just take the value given for $3^7 = 2{,}187$ and multiply by 3 to get 3^8.
   b. $3^8 = 3^7 \times 3 = 2{,}187 \times 3 = 6{,}561$
   c. Since $2^{45} \times 2 = 2^{46}$, just take the value given for $2^{46} = 35{,}184{,}372{,}088{,}832$ and multiply by 2 to get 2^{46}.
   d. $2^{46} = 2^{45} \times 2 = 35{,}184{,}372{,}088{,}832 \times 2 = 70{,}368{,}744{,}177{,}664$

8. a. 17^2 gives us the area of a square with a side length of 17 units.
   b. 101^3 gives us the volume of a cube with an edge length of 101 units.
   c. 2×6^2 gives us the area of two squares with a side length of 6 units.
   d. 4×10^3 gives us the volume of 4 cubes with an edge length of 10 units.

Puzzle Corner:

$2^6 = 64$, $2^5 = 32$, $2^4 = 16$, $2^3 = 8$, $2^2 = 4$, $2^1 = 2$, $2^0 = 1$, so 2^0 equals 1.
$10^6 = 1{,}000{,}000$; $10^5 = 100{,}000$; $10^4 = 10{,}000$; $10^3 = 1{,}000$; $10^2 = 100$; $10^1 = 10$; $10^0 = 1$.
For any number n, $n^0 = 1$ because at each step of the pattern you divide by n.
The next-to-last step will always be $n^1 = n$, and the last step will be $n^0 = n^1/n = n/n = 1$.

Place Value, p. 23

1.

a. The digit "9" is in the millions place. Its value is 9 × a million, or 9 million.							8	9	0	0	2	4	0	0
	trillions period			billions period			millions period			thousands period			ones period	
b. The digit "3" is in the ten thousands place. Its value is 3 × ten thousand, or 30 thousand.				1	4	2	0	0	2	1	3	9	0	0	0
	trillions period			billions period			millions period			thousands period			ones period	
c. The digit "4" is in the ten millions place. Its value is 4 × ten million, or 40 million.			5	0	0	0	0	4	7	0	0	0	2	6	0
	trillions period			billions period			millions period			thousands period			ones period	

2.

a. 56,**8**09 *the hundreds place* value *800*	b. 2**8**7,403,222 the ten millions place value 80,000,000	c. 1**8**,503,200,000,000 trillions place 8,000,000,000,000	d. **8**,493,591,000 billions place 8,000,000,000

3. a. 2,180,027,000
   b. 60,000,000,453,000
   c. 4,050,054,000,009

4.

10^0	1	one
10^1	10	ten
10^2	100	one hundred
10^3	1,000	one thousand
10^4	10,000	ten thousand
10^5	100,000	one hundred thousand
10^6	1,000,000	one million
10^7	10,000,000	ten million
10^8	100,000,000	one hundred million
10^9	1,000,000,000	one billion
10^{10}	10,000,000,000	ten billion
10^{11}	100,000,000,000	one hundred billion
10^{12}	1,000,000,000,000	one trillion

5. a. 80,507
   b. 7,056,060
   c. 7,170,000,000
   d. 604,512,000
   e. 2,305,870,000
   f. 20,160,002

Place Value, cont.

6.

a. $2{,}839 = 2000 + 800 + 30 + 9$ $= 2 \times 10^3 + 8 \times 10^2 + 3 \times 10^1 + 9 \times 10^0$	b. $483 = 400 + 80 + 3$ $= 4 \times 10^2 + 8 \times 10^1 + 3 \times 10^0$

c. $10{,}540 = 10{,}000 + 500 + 40 = 1 \times 10^4 + 5 \times 10^2 + 4 \times 10^1$

d. $407{,}000 = 400{,}000 + 7{,}000 = 4 \times 10^5 + 7 \times 10^3$

e. $500{,}120{,}000 = 500{,}000{,}000 + 100{,}000 + 20{,}000 = 5 \times 10^8 + 1 \times 10^5 + 2 \times 10^4$

f. $4{,}078{,}003 = 4{,}000{,}000 + 70{,}000 + 8{,}000 + 3 = 4 \times 10^6 + 7 \times 10^4 + 8 \times 10^3 + 3 \times 10^0$

7. a. $>$ b. $<$ c. $=$
   d. $>$ e. $<$ f. $<$
   g. $=$ h. $<$ i. $>$

8. a. 1,300,000; 1,400,000; 1,500,000; 1,600,000; 1,700,000; 1,800,000; 1,900,000; 2,000,000; 2,100,000
   b. 724,388; 724,588; 724,788; 724,988; 725,188; 725,388; 725,588; 725,788; 725,988
   c. 15,100,000; 15,500,000; 15,900,000; 16,300,000; 16,700,000; 17,100,000; 17,500,000; 17,900,000; 18,300,000

9.

a. $109{,}000 < 8{,}000{,}000 < 10^8$	b. $9 \times 10^7 < 970{,}000{,}000 < 7 \times 10^9$
c. $54{,}050 < 450{,}055 < 450{,}540$	d. $8 \times 10^6 < 8{,}998{,}998 < 8{,}999{,}000$
e. $45{,}005 < 55{,}400 < 5 \times 10^5$	f. $700{,}000 < 7 \times 10^6 < 6 \times 10^7$

10. a. $10^6 - 10^4 + 50{,}000 = 1{,}040{,}000$
    b. $295{,}209{,}328 - 7{,}399{,}800 - 25{,}906 = 287{,}783{,}622$
    c. $5 \times 10^6 + 456{,}200 + 8 \times 10^9 = 8{,}005{,}456{,}200$

11.

Number	10,000	350,000	1,200,000	74,900	203,060
−1	9,999	349,999	1,199,999	74,899	203,059
− 10	9,990	349,990	1,199,990	74,890	203,050
− 100	9,900	349,900	1,199,900	74,800	202,960
− 1000	9,000	349,000	1,199,000	73,900	202,060

12. a. The estimate of China's population at the end of 2012 was about
    1,354,000,000 + 6,000,000 + 6,000,000 = 1,366,000,000.
    Note: Since the original population figure for 2012 was accurate to the nearest million, the figures you add to it should be rounded to the nearest million, too.

    b. In 2012 China's population was 1,354,000,000 and the USA's population was 315,591,000. The ratio is 1,354,000,000 ÷ 315,591,000 = 4.29..., so the population of China is approximately four times bigger than the population of the USA.

Rounding and Estimating, p. 27

1.

Number	410,987	699,893	12,781,284	3,259,949	1,495,397
...thousand	411,000	700,000	12,781,000	3,260,000	1,495,000
...ten thousand	410,000	700,000	12,780,000	3,260,000	1,500,000
...hundred thousand	400,000	700,000	12,800,000	3,300,000	1,500,000
...million	0	1,000,000	13,000,000	3,000,000	1,000,000

2.

Number	8,419,289,387	12,238,994,038	3,459,994,920	2,203,845,108
...ten million	8,420,000,000	12,240,000,000	3,460,000,000	2,200,000,000
...hundred million	8,400,000,000	12,200,000,000	3,500,000,000	2,200,000,000
...billion	8,000,000,000	12,000,000,000	3,000,000,000	2,000,000,000

3. a. 299,724 ≈ 300,000 b. 1,399,956 ≈ 1,400,000
   c. 698,999,865 ≈ 699,000,000 d. 499,998,325 ≈ 500,000,000

4.

a. 2,384 × 19,384 Estimation: 2,000 × 20,000 = 40,000,000 Exact: 46,211,456 Error of estimation: 6,211,456	b. 345 × 61,852 Estimation: 300 × 62,000 = 18,600,000 Exact: 21,338,940 Error of estimation: 2,738,940
c. 124,012 − 16 × 2,910 Estimation: 124,000 − 10 × 3,000 = 124,000 − 30,000 = 94,000 or 124,000 − 15 × 3,000 = 124,000 − 45,000 = 79,000 Exact: 77,452 Error of estimation: 16,548 or 1,548	d. 25,811 ÷ 487 Estimation: 26,000 ÷ 500 = 52 Exact: 53 Error of estimation: 1

5. 2 × $1.40 × 20 = 56. Janet spends approximately $56 a month for bus fare.

6. 7 × $1.00 + 3 × $11.50 = $7 + $34.50 = $41.50

7. The total monthly savings are $2,417. You can solve this in many different ways. For example:

   Right now, the salaries are a total of 3 × $1,552 + 2 × $1,267 = $4,656 + $2,534 = $7,190.
   When those are cut by 1/10, the savings are $719.
   When the other operating costs, $8,490, are cut by 2/10, the savings are $849 + $849 = $1,698.
   In total, the savings are $1,698 + $719 = $2,417.

Lessons in Problem Solving, p. 30

1. One small carpet costs $55.50 ÷ 5 × 2 = $22.20. Two of them cost $44.40.
   $50 − $44.40 = $5.60 His change was $5.60.

2. The smaller ones hold 0.75 L ÷ 10 × 7 = 0.525 L.
   Four large containers would hold 4 × 0.75 = 3 liters.
   Five small containers would hold 5 × 0.525 = 2.625 liters.
   In total, they hold 3 L + 2.625 L = 5.625 L. So, yes, five liters of soup will fit into four large and five small containers.

3. Converting the 25 kg and 15 kg into grams: 25,000 g ÷ 20 = 1,250 g or 1.25 kg. One bag of bolts weighs 1.25 kg.
   15,000 g ÷ 20 = 750 g or 0.75 kg. One bag of nuts weighs 0.75 kg.
   1.25 kg + 0.75 kg = 2 kg. Together the nuts and bolts weigh 2 kg.

Lessons in Problem Solving, cont.

4. a. 670 ÷ 4 × 3 = 502.5 g. A medium jar holds about 502 grams.
   502.5 ÷ 3 × 2 = 335 g. A small jar holds 335 grams.
   b. The total weight is 670 g + 502 g + 335 g = 1,507 g or 1.507 kg.

5. John had initially $30.60 ÷ 5 × 9 = $55.08.
   Karen had initially $30.60 ÷ 3 × 7 = $71.40.
   $71.40 − $55.08 = $16.32. Karen had $16.32 more than John initially.

6. a. 569 ÷ 43 = 13 R10. So, they need 14 buses. (Of which 13 will be full, and the 14th bus will have 10 people in it.)
   b. The total mileage is 60 miles × 14 buses = 840 miles.
   The total cost is 840 × $2.15 = $1,806.

7. The original price of the first washer is $360 ÷ 9 × 10 = $400.

   The original price of the second washer is $350 ÷ 3 × 5 = $583.33.

   The difference in the two prices is $583.33 − $400 = $183.33.

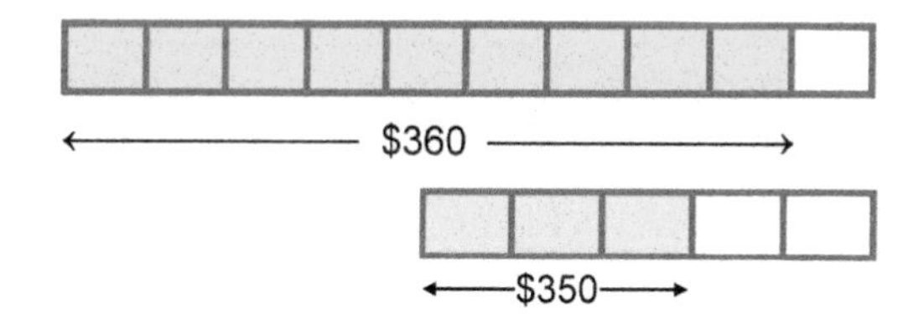

Puzzle corner:

a. Divide 1 by 10, and it gives you 0.1
   Divide 81 by 100, and it gives you 0.81.
   Divide 492 by 1,000, and it gives you 0.492.
   Divide 355 by 100, and it gives you 3.55.
   If the number has tenths, divide it by 10, if it has hundredths divide it by 100; if it has thousandths divide it by 1000, etc.
b. 138 ÷ 100 × 039 ÷ 100 = 0.5382.

Chapter 1 Review, p. 34

1. a. 83 R41   b. 6,735 R45

2. 23,391 ÷ 9 = 2,599 times

3. You will spend 365 × $2.25 = $821.25 in a year on phone calls.

4. 5,000 ÷ 46 = 108 R32. They will need 109 buses.

5. Multiply to estimate, and use 900 km, instead of 880 km. Since 6 × 900 km = 5,400 and 7 × 900 km = 6,300 km, it will take about 6 1/2 hours to travel 5,800 km.

6. $15.90 ÷ 3 × 2 = $10.60. Two boxes of tea bags cost $10.60.

7. a. $5^4 = 625$ d. $100^3 = 1,000,000$
   b. $1^6 = 1$ e. $2^6 = 64$
   c. $30^2 = 900$ f. $3^3 = 27$

8. a. Its area is 400 cm². (One side measures 20 cm.)
   b. Its volume is $(11\text{ m})^3 = 1,331\text{ m}^3$.

9. a. 25^3 gives us the volume of a cube with an edge length of 25 units.
   b. 3×9^2 gives us the area of 3 squares with a side length of 9 units.

10. a. 200,309   b. 28,031,000

11. a. 707,000 < 7,000,000 < 10^7
    b. $5 \times 10^4 < 4 \times 10^5 < 450,000$

12. a. 149,601 ≈ 150,000   b. 2,999,307 ≈ 3,000,000

    c. 597,104,865 ≈ 597,000,000   d. 559,998,000 ≈ 560,000,000

Chapter 2: Expressions and Equations

Terminology for the Four Operations, p. 40

1.

numbers	sum	difference	product	quotient
a. 50 and 2	52	48	100	25
b. 5 and 3	8	2	15	5/3 or 1 2/3

2.

numbers/letters	sum	difference	product	quotient
a. x and 6	$x + 6$	$x - 6$	$6x$	$\frac{x}{6}$
b. z and w	$z + w$	$z - w$	wz	$\frac{z}{w}$

Notice that when variables are multiplied, they are written in alphabetical order. That is why the product of z and w is best written as wz; however zw is acceptable since this has not been taught at this point. See page 55 in the text.

3.

a. the product of 5 and T	$5 - T$	29	**e.** the difference of T and 5
b. the quotient of 5 and T	$T - 5$	$\frac{T}{5}$	**f.** the sum of T and 5
c. the product of 6 and 5	30	5T	**g.** the sum of 5, 15, and 9
d. the quotient of T and 5	$5 \div T$	$T + 5$	**h.** the difference of 5 and T

4.

Statement	Equation
a. The quotient is 5, the divisor is 8, the dividend is 40.	$40 \div 8 = 5$
b. The subtrahend is 30, the difference is 15, and the minuend is 45.	$45 - 30 = 15$
c. The factors are 5, 6, and 8, and the product is 240.	$5 \cdot 6 \cdot 8 = 240$
d. The addends are 7, 8, and 85, and the sum is 100.	$7 + 8 + 85 = 100$

5. a. Answers will vary. Please check the student's work.
   Examples: $5 \cdot 6 \cdot 0 = 0$ or $0 \cdot 2 \cdot 250 = 0$ or $65 \cdot 0 \cdot 12 = 0$
   Any numbers can be used as long as at least one of the three factors is zero.

   b. Answers will vary. For example: $120 \div 40 = 3$ or $120 \div 120 = 1$ or $120 \div 9 = 13$ R3.
   The divisor will need to be more than 8 (because $120 \div 8 = 15$).

   c. Answers will vary. For example: $2 + 2 + 2 + 2 = 8$; $1 + 2 + 3 + 2 = 8$; $1 + 1 + 1 + 1 = 4$; $0 + 0 + 0 + 8 = 8$

6. Subtract the difference from the minuend and you will find the subtrahend.
   a. $56 - \underline{37} = 19$
   b. $4{,}203 - 3650 = 553$

7. Divide the dividend by the quotient, and you will find the divisor..
   a. $\frac{56}{\underline{8}} = 7$
   b. $\frac{535}{\underline{107}} = 5$

8. Divide the product solved by the known factor and you will find the missing factor.
   a. $\underline{12} \cdot 8 = 96$
   b. $7 \cdot \underline{418} = 2{,}926$

Terminology for the Four Operations, cont.

9. a. $7y$ b. $S + 7$ c. $\frac{5s}{8}$ d. $x - 9$
   e. $190 + r$ f. $8d$ g. $9 - x$ h. $n - 14$

10. a. $\frac{2x-1}{3}$ b. $(5 + x)^3$ c. $5(x + 2)$ d. $8(4 + x + 2)$
   e. $2(10 - s)$ f. $\frac{y}{y+4}$ g. $\frac{x+4}{x^2}$

11. a. $7s + 6$ b. $4s - 9$ c. $\frac{5+x}{5-x}$ d. $(6 - x)^2$
   e. $(5 - m)^2$ f. $\frac{w^2}{w-1}$ g. $100 - p^2$ h. $7 - x$
   i. $x^2 + 100$

Puzzle corner:
a. 7, 8, 9. The next higher cube after 504 is $8^3 = 512$, so 8 is the middle number. The others are one higher and one lower: $7 \cdot 8 \cdot 9 = 504$.
b. The sum 621 divided by 3 gives the "average" number 207, so the others are one less and one more: 206, 207, 208.

The Order of Operations, p. 44

1. a. $100 - (50 - 50) = 100$ b. $200 \div (10 + 10) + 5 = 15$ c. $(50 + 50) \cdot 4 - 10 = 390$

2.

a. $\frac{64}{8} \cdot 4 = 32$	b. $\frac{64}{8 \cdot 4} \cdot 2 = 4$	c. $4 \cdot \frac{8}{4} \cdot 2 = 16$
d. $\frac{64}{8 \cdot 4} = 2$	e. $\frac{64}{8} \cdot 4 \cdot 2 = 64$	f. $\frac{4 \cdot 8}{4 \cdot 2} = 4$

3.

a. $150 + 2 \cdot 10 = 170$	b. $5^2 \cdot 2^3 = 200$	c. $3^2 \cdot (150 + 900) \div 3 = 3{,}150$
d. $\frac{12+9}{4+1} = 4\ 1/5$	e. $\frac{5^2}{3^2} = 2\ 7/9$	f. $\frac{2^3}{8} + 10^3 = 1001$
g. $(6 + 6)^2 \cdot (15 - 5)^2 = 14{,}400$	h. $40 + 80 \div 2 \cdot 4 - 15 = 185$	i. $\frac{7^2}{7} \cdot 7 = 49$

4. a. $20{,}000 - 7 \cdot 500 = 16{,}500$
   b. $6 \cdot 70 + 5 \cdot 120 = 1{,}020$

5. a. $2^7 + 5^3 = 128 + 125 = 253$
   b. $5 \cdot 100^3 - 2 \cdot 10^5 = 5{,}000{,}000 - 200{,}000 = 4{,}800{,}000$

6. a. $(3.2 + 5.3) \cdot 2 = 17$ or $2 \cdot (3.2 + 5.3) = 17$ b. $(190 - 50) \div 5 = 28$
   c. $100 - (40 - 5) = 65$ d. $(2 \cdot 5)^3 = 1{,}000$
   e. $\frac{5}{3^3} = 5/27$ f. $2(10 - 4) + 3(5 + 8) = 51$
   g. $\frac{61 - 30}{5^2} = 31/25 = 1\ 6/25$ h. $6^2 - 7 = 29$

7. a. $5 \cdot 10¢ + 15 \cdot 1¢ + 2 \cdot 25¢ + 7 \cdot 50¢ = 465¢$
   b. $(64 - 15) \cdot 2 = 98$. Henry has 98 marbles.
   c. $5 \cdot 20 - 9 \cdot 2 = 82$. The colored area is 82 square units.
   d. $5 \cdot 5 - 2 \cdot 2 = 21$. The colored area is 21 square units.

Multiplying and Dividing in Parts, p. 47

1.

a. $7 \cdot 99 = 7 \cdot (100 - 1)$ $= 700 - 7 = 693$	b. $4 \cdot 999 = 4 \cdot (1{,}000 - 1)$ $= 4{,}000 - 4 = 3{,}996$
c. $5 \cdot 104 = 5 \cdot (100 + 4)$ $= 500 + 20 = 520$	d. $5 \cdot 998 = 5 \cdot (1{,}000 - 2)$ $= 5{,}000 - 10 = 4{,}990$
e. $6 \cdot 98 = 6 \cdot (100 - 2)$ $= 600 - 12 = 588$	f. $7 \cdot 2030 = 7 \cdot (2{,}000 + 30)$ $= 14{,}000 + 210 = 14{,}210$

2.

a. 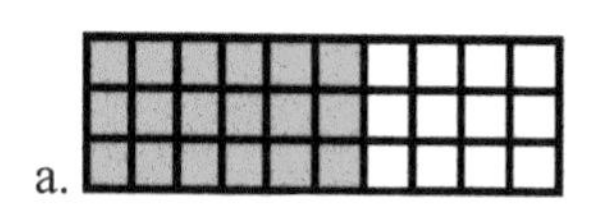

Total area: $3 \cdot (6 + 4)$
The areas of the two rectangles:
$3 \cdot 6$ and $3 \cdot 4$

b.

Total area: $4 \cdot (5 + 4)$
The areas of the two rectangles:
$4 \cdot 5$ and $4 \cdot 4$

c. 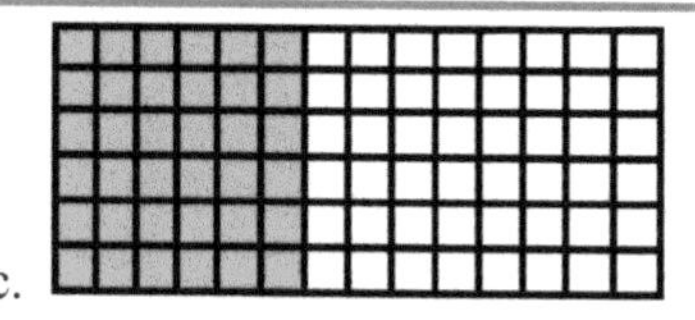

Total area: $6 \cdot (6 + 8)$
The areas of the two rectangles:
$6 \cdot 6$ and $6 \cdot 8$

d. 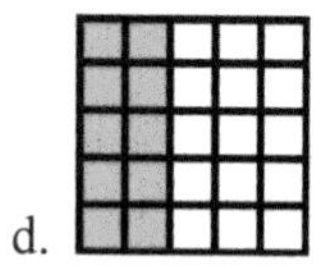

Total area: $5 \cdot (2 + 3)$
The areas of the two rectangles:
$5 \cdot 2$ and $5 \cdot 3$

3. a. 80 is the partial product of 10 times 8. (10 from 16 and 8 from 78)

700 is the partial product of 10 times 70 (10 from 16 and 70 from 78).

b.

```
     5 6
   x 8 4
     2 4
   2 0 0
   4 8 0
 4 0 0 0
 4 7 0 4
```

c.

```
     1 7
   x 9 5
     3 5
     5 0
   6 3 0
   9 0 0
 1 6 1 5
```

4.

a. $29 \cdot 17$

200	90
140	63

$29 \cdot 17 = 20 \cdot 10 + 20 \cdot 7$
$+ 9 \cdot 10 + 9 \cdot 7$
$= 200 + 140 + 90 + 63 = 493$

b. $75 \cdot 36$

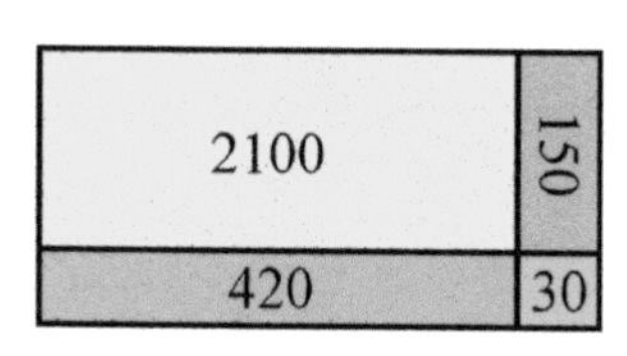

$75 \cdot 36 = 70 \cdot 30 + 70 \cdot 6$
$+ 5 \cdot 30 + 5 \cdot 6$
$= 2{,}100 + 420 + 150 + 30 = 2{,}700$

5.

a. $\frac{80}{2} + \frac{12}{2} = 40 + 6 = 46$	b. $\frac{350}{5} + \frac{15}{5} = 70 + 3 = 73$	c. $\frac{400}{4} - \frac{12}{4} = 100 - 3 = 97$
d. $\frac{9{,}300}{3} - \frac{60}{3} = 3{,}100 - 20 = 3{,}080$	e. $\frac{350}{7} + \frac{21}{7} - \frac{7}{7} = 50 + 3 - 1 = 52$	f. $\frac{900}{9} - \frac{18}{9} = 100 - 2 = 98$
g. $\frac{22 \text{ ft}}{2} + \frac{9 \text{ in}}{2} = 11$ ft 4.5 in	h. $\frac{40 \text{ kg}}{5} + \frac{750 \text{ g}}{5} = 8$ kg + 150 g	i. $\frac{12 \text{ L}}{4} + \frac{600 \text{ ml}}{4} = 3$ L 150 ml

Multiplying and Dividing in Parts, cont.

6. a. 206 b. 203 c. 103 d. 201 e. 502

7.

a. $\frac{15}{5} + \frac{4}{5} = 3\frac{4}{5}$	b. $\frac{44}{11} + \frac{7}{11} = 4\frac{7}{11}$
c. $\frac{6}{7} + \frac{70}{7} = 10\frac{6}{7}$	d. $\frac{420}{6} + \frac{2}{6} = 70\frac{2}{6}$
e. $\frac{240}{4} + \frac{12}{4} + \frac{3}{4} = 60 + 3 + \frac{3}{4} = 63\frac{3}{4}$	f. $\frac{2}{9} + \frac{36}{9} + \frac{270}{9} = 4 + 30 + \frac{2}{9} = 34\frac{2}{9}$

8. a. 100 3/4 b. 303 2/3 c. 1,004 4/5
d. 20 1/4 e. 42 1/3 f. 60 5/6

9. a. 20 kg 9 3/10 g b. 3 m 2/5 cm c. 1 ft 7/10 in
d. 9 ft 1 4/5 in e. 6 m 2.5 cm f. 16 gal 1 1/3 qt or 16 1/3 gal

10. One way: Two quarts and 10 ounces equal 64 oz + 10 oz = 74 oz. Then, 74 oz ÷ 4 = 18 1/2 ounces per person.

Another way: Two quarts divided among 4 people is half a quart or one pint (16 oz) for each. Ten ounces divided among 4 people is 10 oz/4 = 2 ½ oz per person. So each of the four people gets 1 pint plus 2 ½ oz, or 16 + 2 ½ = 18 ½ ounces.

11. a. 7 ÷ 14 = 1/2 b. 7 ÷ 21 = 1/3 c. 80 ÷ 11 = 7 3/11
d. 6/8 + 3 + 30 = 33 6/8 e. 117 ÷ 4 = 29 1/4 f. 100 ÷ 30 = 3 1/3

Puzzle corner:

a. $\frac{250 - 3}{10} = 25 - \frac{3}{10}$ b. $\frac{11 - 3}{5} = 2\frac{1}{5} - \frac{3}{5}$

Expressions, p. 51

1.

a. 80/9 = 8 8/9	b. 27 − 15 = 12	c. 10,000 ÷ 100 = 100
d. 1,000 − 600 = 400	e. (1/9) × 81 − 4 = 9 − 4 = 5	f. 6,000 ÷ 500 = 12

2.

Variable	Expression $100 - x^2$	Value
$x = 3$	$100 - 3^2 = 100 - 9$	91
$x = 4$	$100 - 4^2 = 100 - 16$	84
$x = 5$	$100 - 5^2 = 100 - 25$	75
$x = 6$	$100 - 6^2 = 100 - 36$	64
$x = 7$	$100 - 7^2 = 100 - 49$	51

3. a. $2 \times 5 + 18 = 10 + 18 = 28$ b. $\frac{35}{5} \times 13 = 7 \times 13 = 91$
c. $5 \times 9 = 45$ d. $\frac{3}{5} \times 25 = 15$

4. a. 80 − 14 − 5 = 61 b. 80 − (14 − 5) = 71
c. 80 + 14 + 5 = 99 d. 80 − (14 + 5) = 61

5. a. 4a and 4d had the same value.
b. Yes. The values for 4a and 4d are the same even using different numbers.
c. Yes, they are equivalent expressions.

Expressions, cont.

6. a. $30(s - 300)$

   b. $\dfrac{35 + x}{7}$

   c. $y - \dfrac{200}{40}$

7. a. $30(1200 - 300) = 27{,}000$

   b. $\dfrac{35 + 42}{7} = 11$

   c. $800 - \dfrac{200}{40} = 795$

8. a. $V = (4\text{ cm})^3 = 64\text{ cm}^3$
   b. $V = (½\text{ in})^3 = 1/8\text{ in}^3$

9. Cost $= \dfrac{\$3.25 \times 380}{22} = \$56.14$

10. a. $10p$ b. $(1/5)S$ or $S/5$ c. $2m + 1.50n$ d. $p + q$ e. $\dfrac{p + q}{2}$

11. a. The first 3 expressions are equivalent.
    b. The first and third expressions are equivalent ($2x/6$ and $x/3$).

Writing and Simplifying Expressions 1: Length and Perimeter, p. 54

1. a. $5x$ b. $2x + 2y$
   c. $3n + m$ d. $3z + 8$
   e. $2q + x + 3$ f. $2z + 2x + y + 9$

2. a. $2x + 8$ b. $2x + 2y$
   c. $6s$ d. $4y$
   e. $a + b + 7$ f. $2s + 2t + 3$

3. a. $3x + x = 4x$ b. $2z + 7 + 12 = 2z + 19$
   c. $4p - 2p = 2p$ d. $4x + 11 - 2x = 2x + 11$

4.

a. $5c + 2$	b. $3p + 2r$
c. $3x + 7$	d. $2x + 3z + 4$ *or* $3z + 2x + 4$
e. $3m + q + s$	f. $2y + 13$
g. $7c$	h. $6p$
i. $5d$	j. $9x$
k. $4x + 5$ *or* $5 + 4x$	l. $9a + 2 - 7x$ *or* $9a - 7x + 2$

5.

a. c^3	b. $4x^2$
c. $30x$	d. $2xz$
e. $45ab$	f. $4y + 8$
g. $c^2 + 16$	h. $32r - 14$
i. $280wx$	j. $3p^3$
k. $60w^4$	l. $r^2 - 9$

More on Writing and Simplifying Expressions, p. 57

1.

Expression	the terms in it	coefficient(s)	Constants
$4a + 5b$	$4a$ and $5b$	4, 5	none
$300y$	$300y$	300	none
$11x + 5$	$11x$ and 5	11	5
$x + 12y + 9$	x, $12y$, and 9	1, 12	9
$p \cdot 9$	$p \cdot 9$	9	none
$8x^4y^3 + 10$	$8x^4y^3$ and 10	8	10
$\frac{11}{26}p$	$\frac{11}{26}p$	$\frac{11}{26}$	none

2. a. $x^2 + \frac{1}{2}$

   b. $2a + 6b + 7$ or $6a + 2b + 7$

3.

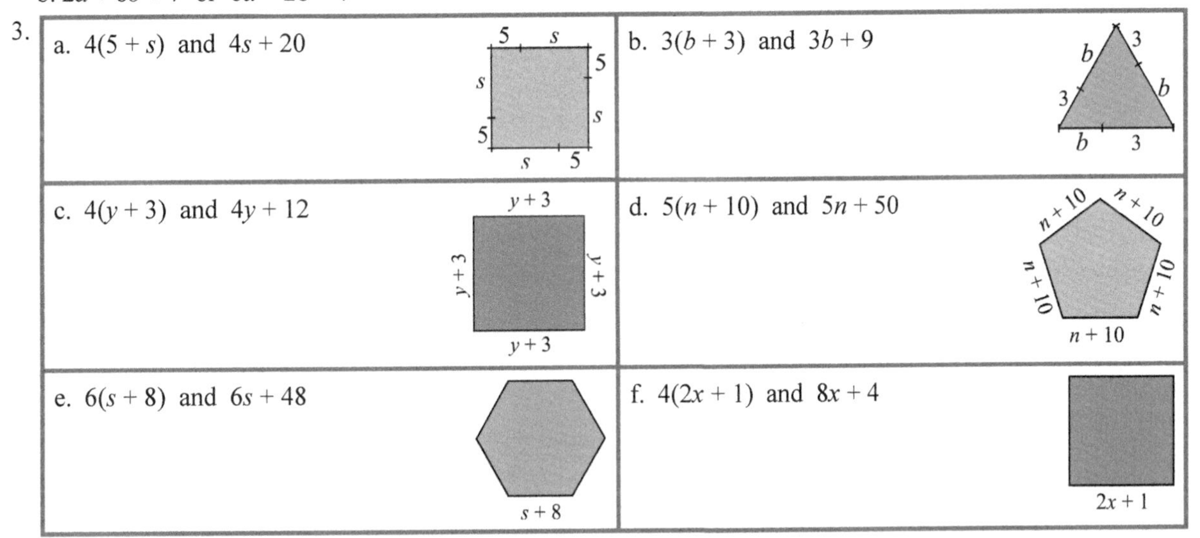

a. $4(5 + s)$ and $4s + 20$	b. $3(b + 3)$ and $3b + 9$
c. $4(y + 3)$ and $4y + 12$	d. $5(n + 10)$ and $5n + 50$
e. $6(s + 8)$ and $6s + 48$	f. $4(2x + 1)$ and $8x + 4$

4. a. $4(a + 20)$ b. $5m$
   c. $20p$ d. $v - 15$
   e. $p - \$5$ f. $5(p - \$3)$

5.

Expression	Like terms, if any	Constants
$15x + 12z + 9z$	$12z$ and $9z$	none
$10 + 10y + 30y$	$10y$ and $30y$	10
$p \cdot 9 + 2$	none	2
$8a - 2a + 10 + b + 7b$	$8a$ and $2a$; b and $7b$	10
$8y + 7x + 6 + 15y - 2x$	$8y$ and $15y$; $7x$ and $2x$	6

6. a. $9x + 8$ b. $13z^2$
   c. $3m + 12n$ d. $5x + 23y + 6$
   e. $10m^2 + 9$ f. not possible to simplify

7. 5.0 gallons equals 18.9 liters.

8. a. $3 + 2(½) = 4$
   b. $2(3 + ½) = 7$

9. a. $s + s + s$, $3s$, and $2s + s$
   b. $2y + 2x$ and $y + y + x + x$

Writing and Simplifying Expressions 2: Area, p. 60

Teaching box: Its perimeter is $2(l + w)$.

1. a. $A = 3x \cdot 4x = 12 \cdot x \cdot x = 12x^2$
   b. $A = 5y \cdot 3y = 15 \cdot y \cdot y = 15y^2$
   c. $A = 5x \cdot x = 5 \cdot x \cdot x = 5x^2$
   d. $A = 2b \cdot 7b = 14b \cdot b = 14b^2$

2. a. $A = 12x^2$ $P = 16x$
   b. $A = 3x^2$ $P = 8x$
   c. $A = 9s^2$ $P = 12s$
   d. $A = 8a^2$ $P = 12a$

3. a. $A = 28c^2$ $P = 22c$
   b. $A = 80x^2$ $P = 36x$

4. a. $A = 14x^2$ $P = 18x$
   b. $A = 19x^2$ $P = 20x$

5. a. $A = 19x^2$ $P = 26x$
   b. $A = 24x^2$ $P = 22x$
   c. $A = 40b^2$ $P = 32b$
   d. $A = 43b^2$ $P = 28b$
   e. $A = 51a^2$ $P = 36a$
   f. $A = 51a^2$ $P = 36a$

6. a. $A = 20x^2$
   b. $A = 500 \text{ cm}^2$
   c. $A = 2{,}000 \text{ cm}^2$
   d. No. It is quadruple (four times as much).

7. a. $A = 51a^2$ $P = 36a$
   b. 72 in
   c. 144 in
   d. yes
   e. $51 \cdot 2^2 = 204$ sq. in. and $51 \cdot 4^2 = 816$ sq. in. No, the area was quadruple (four times as much).

8. a. $3b$ b. $6a$

9. a. $6c$ b. $6b$

10. The length of each side is $2y$.

11. The other side is $8s$.

12. a. $81r^2$
    b. $40a$
    c. Ava's age is $S - 3$.
    d. The more expensive shoes cost p + \$10. e. $5(p + \$10)$

13. a. $M = \dfrac{24.0}{1.60934} \approx 14.9$ miles
    b. You would multiply the number of miles by 1.60934.

The Distributive Property, p. 65

1.

a. $3(90 + 5) = 3 \cdot 90 + 3 \cdot 5 = 270 + 15 = 285$	b. $7(50 + 6) = 7 \cdot 50 + 7 \cdot 6 = 350 + 42 = 392$
c. $4(a + b) = 4 \cdot a + 4 \cdot b = 4a + 4b$	d. $2(x + 6) = 2 \cdot x + 2 \cdot 6 = 2x + 12$
e. $7(y + 3) = 7 \cdot y + 7 \cdot 3 = 7y + 21$	f. $10(s + 4) = 10 \cdot s + 10 \cdot 4 = 10s + 40$
g. $s(6 + x) = s \cdot 6 + s \cdot x = 6s + sx$	h. $x(y + 3) = x \cdot y + x \cdot 3 = xy + 3x$
i. $8(5 + b) = 8 \cdot 5 + 8 \cdot b = 40 + 8b$	j. $9(5 + c) = 9 \cdot 5 + 9 \cdot c = 45 + 9c$

2.

a. $3(a + b + 5) = 3a + 3b + 15$	b. $8(5 + y + r) = 40 + 8y + 8r$
c. $4(s + 5 + 8) = 4s + 52$	d. $3(10 + c + d + 2) = 36 + 3c + 3d$

3.

a. $2(3x + 5) = 6x + 10$	b. $7(7a + 6) = 49a + 42$
c. $5(4a + 8b) = 20a + 40b$	d. $2(4x + 3y) = 8x + 6y$
e. $3(9 + 10z) = 27 + 30z$	f. $6(3x + 4 + 2y) = 18x + 24 + 12y$
g. $11(2c + 7a) = 22c + 77a$	h. $8(5 + 2a + 3b) = 40 + 16a + 24b$

4.

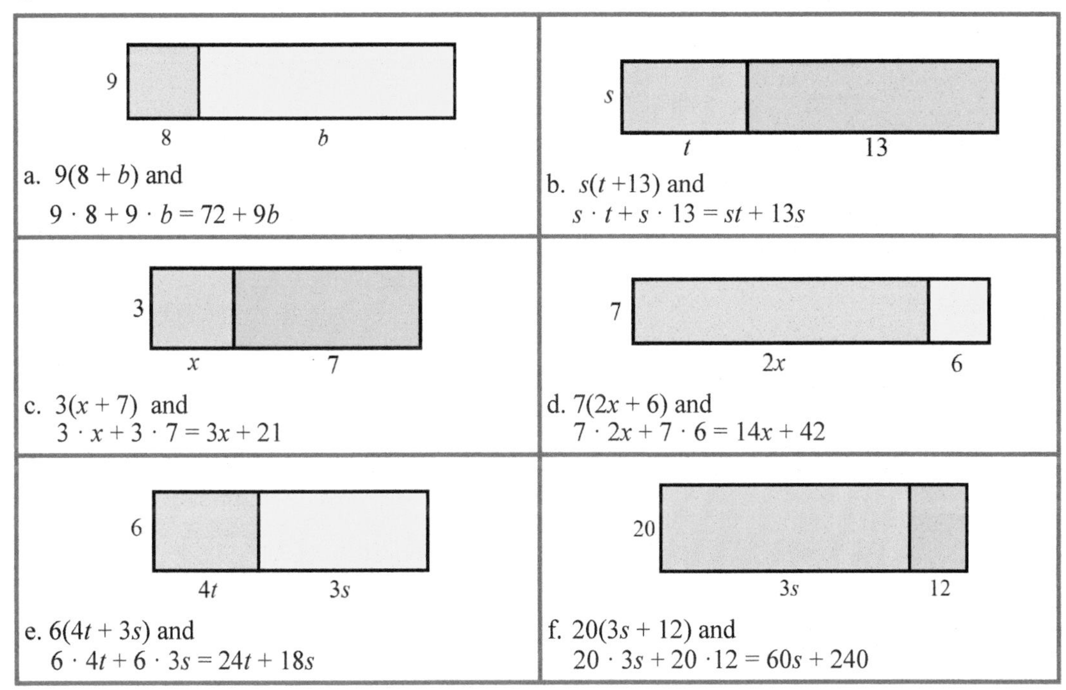

a. $9(8 + b)$ and $9 \cdot 8 + 9 \cdot b = 72 + 9b$	b. $s(t + 13)$ and $s \cdot t + s \cdot 13 = st + 13s$
c. $3(x + 7)$ and $3 \cdot x + 3 \cdot 7 = 3x + 21$	d. $7(2x + 6)$ and $7 \cdot 2x + 7 \cdot 6 = 14x + 42$
e. $6(4t + 3s)$ and $6 \cdot 4t + 6 \cdot 3s = 24t + 18s$	f. $20(3s + 12)$ and $20 \cdot 3s + 20 \cdot 12 = 60s + 240$

5. a. 3 b. 7 c. 9 d. z e. 6 f. 6

6. a. 6 b. 3 c. 2 d. 3

7. a. 3 b. 7 c. 2 d. 5 e. $2x$ f. $3x$ g. $4y$ h. $5t + s$

8. $7(2x + 3) = 14x + 21$

9. One side would be $3x + 1$.

10.

a. $6x + 6 = 6(x + 1)$	b. $8y + 16 = 8(y + 2)$
c. $15x + 45 = 15(x + 3)$	d. $4w + 40 = 4(w + 10)$
e. $6x + 30 = 6(x + 5)$	f. $8x + 16y + 48 = 8(x + 2y + 6)$

11.

a. $8x + 4 = 4(2x + 1)$	b. $15x + 10 = 5(3x + 2)$
c. $24y + 8 = 8(3y + 1)$ or $2(12y + 4)$ or $4(6y + 2)$	d. $6x + 3 = 3(2x + 1)$
e. $42y + 14 = 7(6y + 2)$ or $2(21y + 7)$	f. $32x + 24 = 8(4x + 3)$ or $2(16x + 12)$
g. $27y + 9 = 3(9y + 3)$ or $9(3y + 1)$	h. $55x + 22 = 11(5x + 2)$
i. $36y + 12 = 3(12y + 4)$ or $4(9y + 3)$ or $2(18y + 6)$ or $6(6y + 2)$	j. $36x + 9z + 27 = 3(12x + 3z + 9)$ or $9(4x + 1z + 3)$

The Distributive Property, cont.

12. Its side is $12x + 4$.

Puzzle corner: $10,300 ÷ 600 = $17.17 per item.

Equations, p. 69

1. a. equation   b. expression   c. equation   d. expression   e. expression

2. b. Equation: $100 - x = 35$ Solution: $x = 65$
   c. Equation: $3x = 63$ Solution: $x = 21$
   d. Equation: $x \div 7 = 12$ Solution: $x = 84$
   e. Equation: $x - 19 = 394$ Solution: $x = 413$
   f. Equation: $60 \div x = 12$ Solution: $x = 5$

3. a. No.
   b. Yes.

4. a. The roots are 3 and 6: $3^2 + 18 - 9 \cdot 3 = 0$ and $6^2 + 18 - 9 \cdot 6 = 0$.
   b. The root is 5: $3 \cdot 5 - 5 = 2 \cdot 5$.

5. The number 2 makes the equation true: $\frac{2+6}{2+2} = 2$

6. Answers will vary. Please check the students' work. For example: $\frac{6x}{2} = 33$ or $x^2 - 88 = 33$ or $x + 20 = 31$.

7.

Item	p	$(4/5)p$
a bottle of water	$1	$0.80
a pair of socks	$2	$1.60
a sandwich	$5	$4.00
a ball	$10	$8.00
a toy	$45	$36.00

8. a. $x = 15$
   b. $y = 12$
   c. $z = 35$

9.

a. $54 + x = 990 \quad \vert -54$ $54 + x - 54 = 990 - 54$ $x = 936$	b. $x + 5.6 = 12.9 \quad \vert -5.6$ $x + 5.6 - 5.6 = 12.9 - 5.6$ $x = 7.3$
c. $x - 120 = 137 \quad \vert +120$ $x - 120 + 120 = 137 + 120$ $x = 257$	d. $w - 98 = 89 \quad \vert +98$ $w - 98 + 98 = 89 + 98$ $w = 187$
e. $156 + s = 1{,}082 \quad \vert -156$ $156 + s - 156 = 1{,}082 - 156$ $s = 926$	f. $t + 77 = 208 \quad \vert -77$ $t + 77 - 77 = 208 - 77$ $t = 131$

Equations, cont.

10.

a. $5x = 350$ $\quad\mid \div 5$ $\frac{5x}{5} = \frac{350}{5}$ $x = 70$	b. $10x = 17$ $\quad\mid \div 10$ $\frac{10x}{10} = \frac{17}{10}$ $x = 1.7$
c. $7a = 2.8$ $\quad\mid \div 7$ $\frac{7a}{7} = \frac{2.8}{7}$ $a = 0.4$	d. $\frac{x}{51} = 4$ $\quad\mid \times 51$ $\frac{x}{51} \cdot 51 = 4 \cdot 51$ $x = 204$
e. $\frac{x}{9} = 60$ $\quad\mid \times 9$ $\frac{x}{9} \cdot 9 = 60 \cdot 9$ $x = 540$	f. $\frac{x}{100} = 1.2$ $\quad\mid \times 100$ $\frac{x}{100} \cdot 100 = 1.2 \cdot 100$ $x = 120$

More Equations, p. 73

1.

a. $2y + 5y = 49$ $7y = 49$ $\quad\mid \div 7$ $y = 7$	b. $10x - 8x = 42$ $2x = 42$ $\quad\mid \div 2$ $x = 21$	c. $7a + 2a - 5a = 52$ $4a = 52$ $\quad\mid \div 4$ $a = 13$

2.

a. $y \div 400 = 6 + 2$ $y \div 400 = 8$ $\quad\mid \times 400$ $y = 3{,}200$	b. $z - 220 = 3 \cdot 100$ $z - 220 = 300$ $\quad\mid + 220$ $z = 520$	c. $8x = 501 + 59$ $8x = 560$ $\quad\mid \div 8$ $x = 70$

3.

a. $2x + 3x = 29 - 14$ $5x = 15$ $\quad\mid \div 5$ $x = 3$	b. $7c - c = 3 \cdot 80$ $6c = 240$ $\quad\mid \div 6$ $c = 40$	c. $14x - 6x + 2x = 5 \cdot 40$ $10x = 200$ $\quad\mid \div 10$ $x = 20$

4. a. $9y + 10$ b. $4a^2$
   c. $28s + 9$ d. $3x + 10y$
   e. $7mn + 12$ f. $14w + 3x + 15y$

$7 \times c - c$ $\qquad 7c - 1c = 3 \cdot 80$

5. a. $5n = 485$; $n = 97$. There are 97 nickels.
   b. $25b = \$112.50$; $b = \$4.50$. One bucket costs \$4.50.
   c. $89 - 16\frac{1}{2} = a$; $a = 72\frac{1}{2}$. Ann will be 72 ½ years old when Elizabeth is 89.
   d. $s^2 = 169$; $s = 13$. One side of the square is 13 feet long.

More Equations, cont.

6.

a.
$$\begin{aligned} 2x + 5 &= 27 && \mid -5 \\ 2x + 5 - 5 &= 27 - 5 \\ 2x &= 22 && \mid \div 2 \\ x &= 11 \end{aligned}$$

b.
$$\begin{aligned} 3x - 8 &= 34 && \mid +8 \\ 3x - 8 + 8 &= 34 + 8 \\ 3x &= 42 && \mid \div 3 \\ x &= 14 \end{aligned}$$

c.
$$\begin{aligned} 7x + 5 &= 54 && \mid -5 \\ 7x + 5 - 5 &= 54 - 5 \\ 7x &= 49 && \mid \div 7 \\ x &= 7 \end{aligned}$$

7.

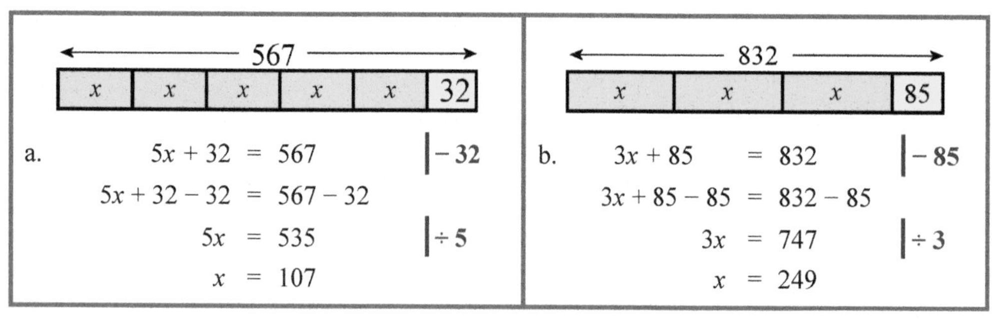

a.
$$\begin{aligned} 5x + 32 &= 567 && \mid -32 \\ 5x + 32 - 32 &= 567 - 32 \\ 5x &= 535 && \mid \div 5 \\ x &= 107 \end{aligned}$$

b.
$$\begin{aligned} 3x + 85 &= 832 && \mid -85 \\ 3x + 85 - 85 &= 832 - 85 \\ 3x &= 747 && \mid \div 3 \\ x &= 249 \end{aligned}$$

8.

a.
$$\begin{aligned} 2(x + 5) &= 24 \\ 2x + 10 &= 24 && \mid -10 \\ 2x &= 14 && \mid \div 2 \\ x &= 7 \end{aligned}$$

OR

a.
$$\begin{aligned} 2(x + 5) &= 24 && \mid \div 2 \\ x + 5 &= 12 && \mid -5 \\ x &= 7 \end{aligned}$$

b.
$$\begin{aligned} 3(x - 4) &= 36 \\ 3x - 12 &= 36 && \mid +12 \\ 3x &= 48 && \mid \div 3 \\ x &= 16 \end{aligned}$$

OR

b.
$$\begin{aligned} 3(x - 4) &= 36 && \mid \div 3 \\ x - 4 &= 12 && \mid +4 \\ x &= 16 \end{aligned}$$

c.
$$\begin{aligned} 7(x + 8) &= 63 \\ 7x + 56 &= 63 && \mid -56 \\ 7x &= 7 && \mid \div 7 \\ x &= 1 \end{aligned}$$

OR

c.
$$\begin{aligned} 7(x + 8) &= 63 && \mid \div 7 \\ x + 8 &= 9 && \mid -8 \\ x &= 1 \end{aligned}$$

d.
$$\begin{aligned} 5(2x + 1) &= 45 \\ 10x + 5 &= 45 && \mid -5 \\ 10x &= 40 && \mid \div 10 \\ x &= 4 \end{aligned}$$

OR

d.
$$\begin{aligned} 5(2x + 1) &= 45 && \mid \div 5 \\ 2x + 1 &= 9 && \mid -1 \\ 2x &= 8 && \mid \div 2 \\ x &= 4 \end{aligned}$$

e.
$$\begin{aligned} 3(4x - 3) &= 51 \\ 12x - 9 &= 51 && \mid +9 \\ 12x &= 60 && \mid \div 12 \\ x &= 5 \end{aligned}$$

OR

e.
$$\begin{aligned} 3(4x - 3) &= 51 && \mid \div 5 \\ 4x - 3 &= 17 && \mid +3 \\ 4x &= 20 && \mid \div 4 \\ x &= 5 \end{aligned}$$

f.
$$\begin{aligned} 3(2x + 7) &= 63 \\ 6x + 21 &= 63 && \mid -21 \\ 6x &= 42 && \mid \div 6 \\ x &= 7 \end{aligned}$$

OR

f.
$$\begin{aligned} 3(2x + 7) &= 63 && \mid \div 3 \\ 2x + 7 &= 21 && \mid -7 \\ 2x &= 14 && \mid \div 2 \\ x &= 7 \end{aligned}$$

9.

p	*New Price*
$24,200	$16,251.17

p	*New Price*
$17,500	$11,751.88

p	*New Price*
$36,400	$24,443.91

10. a. $8(b + 3)$
 b. $t - 3s + 5$

Inequalities, p. 77

1. The variables (letters) that students choose will vary.
   a. $S < \$40$ b. $a \geq 18$ c. $g > 10$ d. $w \geq 12$ e. $p \leq 5$ f. $c \leq 12$

2. Answers will vary. Please check the students' work. Examples:

   a. You have to be less than ten to join the club.
   b. This game is suitable for age 7 or older.
   c. It will take more than 200 days to make the trip. OR The price has to be set at more than $200.
   d. You have to weigh not more than 89 pounds to jump on the trampoline.
   e. He wants to buy at least 2,000 pounds of hay.

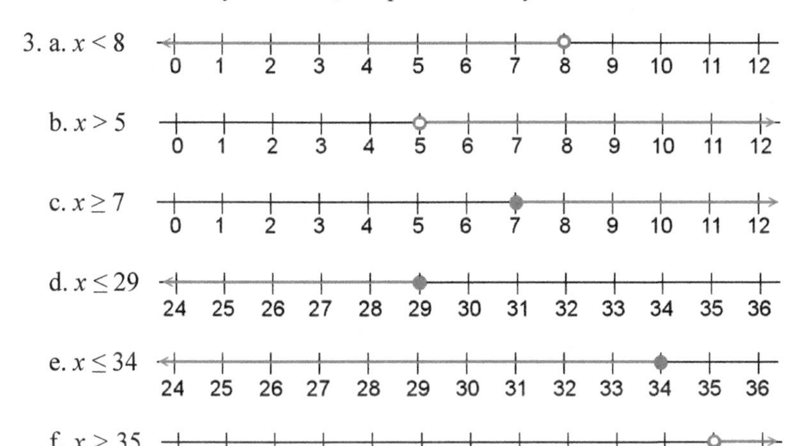

4.

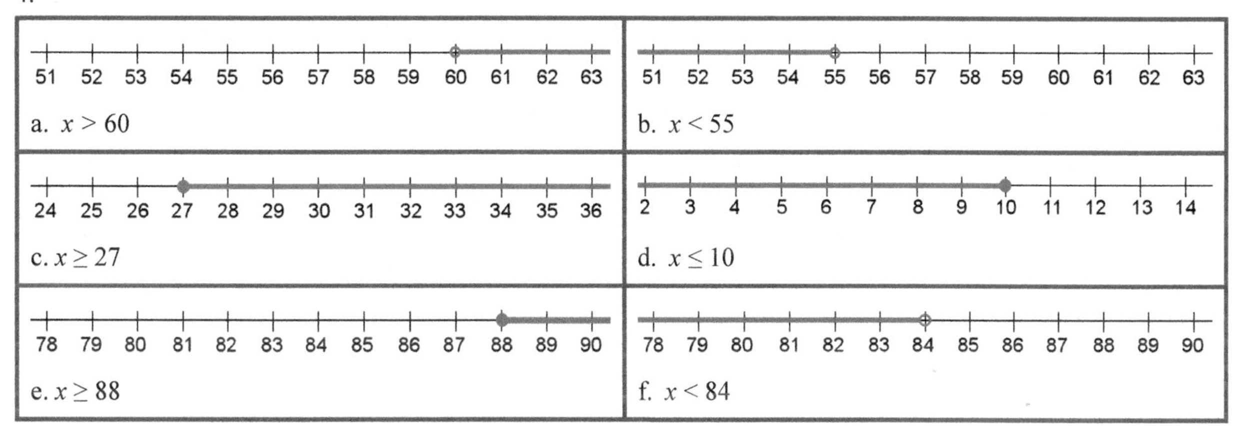

5. a. 15, 18 b. 1, 4, 7, 10, 13 c. 9, 10, 11
   d. 12, 13, 14 e. 6, 7, 8 f. 8, 10

6. a. 1, 2, 3, 4, 5
   b. 2, 4
   c. 18, 20, 22, 24, ...
   d. 2, 4, 6, 8, 10, 12

7.

a. $2y < 48 \quad \vert \div 2$ $y < 24$	b. $x + 8 > 42 \quad \vert -8$ $x > 34$	c. $b - 5 \geq 50 \quad \vert +5$ $b \geq 55$
d. $y - 22 \leq 9 \quad \vert +22$ $y \leq 31$	e. $x + 5.4 < 10.9 \quad \vert -5.4$ $x < 5.5$	f. $20r \leq 900 \quad \vert \div 20$ $r \leq 45$

8.

a. $3x < 30$

$x < 10$

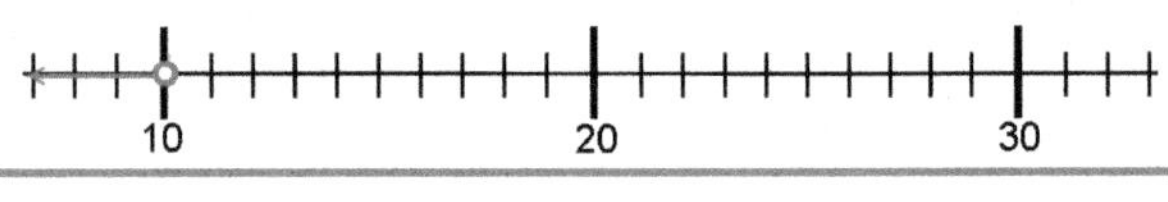

b. $x + 5 > 53$

$x > 48$

c. $x - 15 \leq 37$

$x \leq 52$

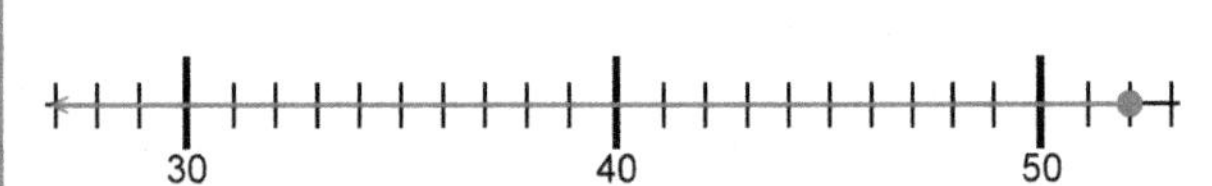

d. $5z \geq 95$

$z \geq 19$

9. a. **(i)**
   b. **(i)**

Using Two Variables, p. 81

1. $y = x + 2$

x	0	1	2	3	4	5	6	7	8
y	2	3	4	5	6	7	8	9	10

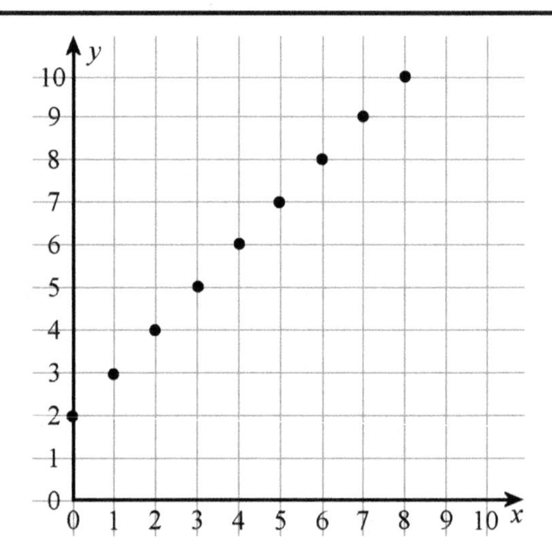

2. $y = 8 - x$

x	0	1	2	3	4	5	6	7	8
y	8	7	6	5	4	3	2	1	0

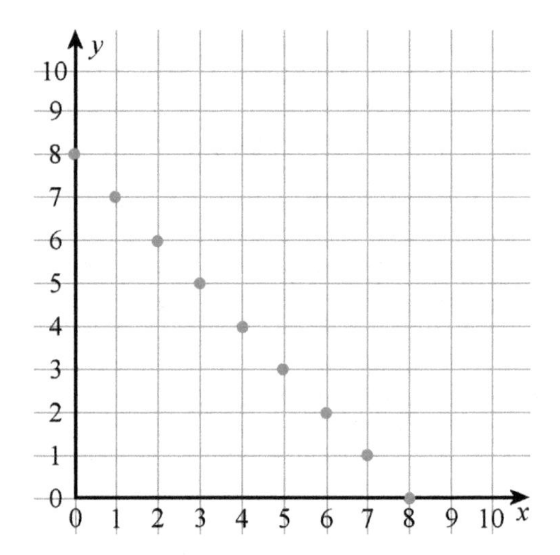

3. $y = 2x - 1$

x	1	2	3	4	5	6
y	1	3	5	7	9	11

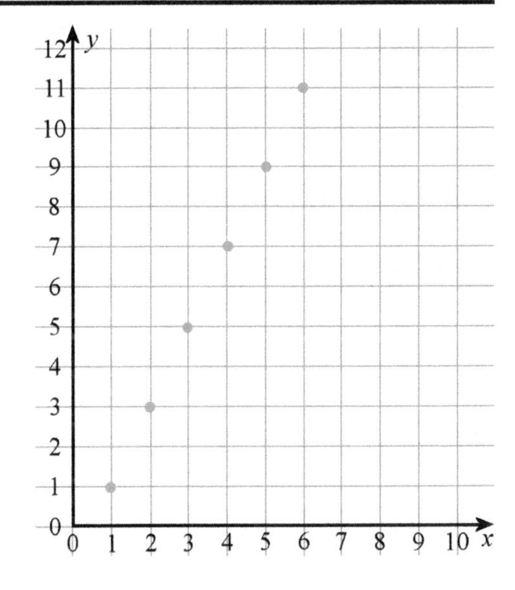

4.

t (hours)	0	1	2	3	4	5	6
d (miles)	0	50	100	150	200	250	300

c. $d = 50t$
d. t is the independent variable.

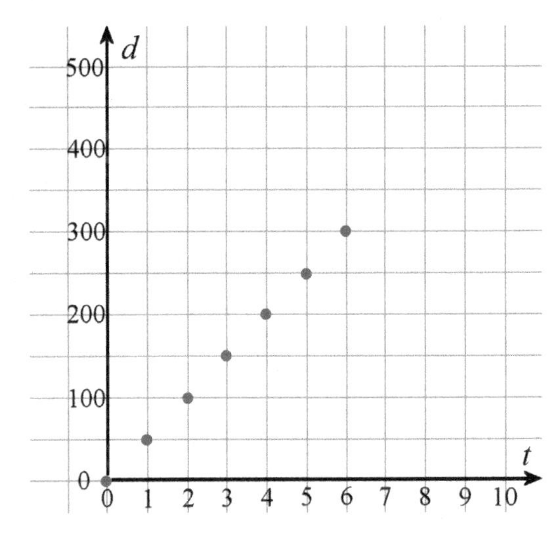

5.

t (sec)	0	1	2	3	4	5	6	7	8	9
V (liters)	0	$\frac{1}{3}$	$\frac{2}{3}$	1	$1\frac{1}{3}$	$1\frac{2}{3}$	2	$2\frac{1}{3}$	$2\frac{2}{3}$	3

c. $V = \frac{1}{3}t$
d. t is the independent variable.

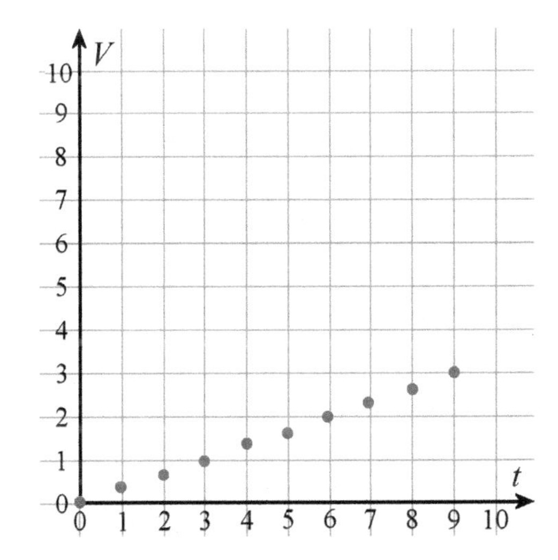

6.

X	3	4	5	6	7	8	9
Y	9	8	7	6	5	4	3

b. $12 - X = Y$

c. x is the independent variable.

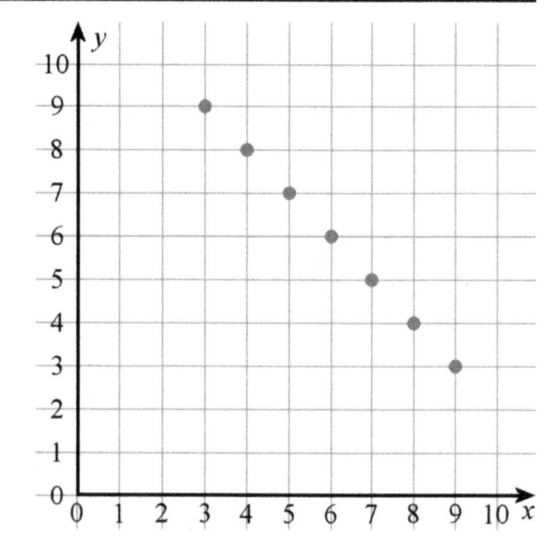

7.

A	5	10	15	20	25	30
H	215	210	205	200	195	190

A	35	40	45	50	55	60
H	185	180	175	170	165	160

c. H is the dependent variable.

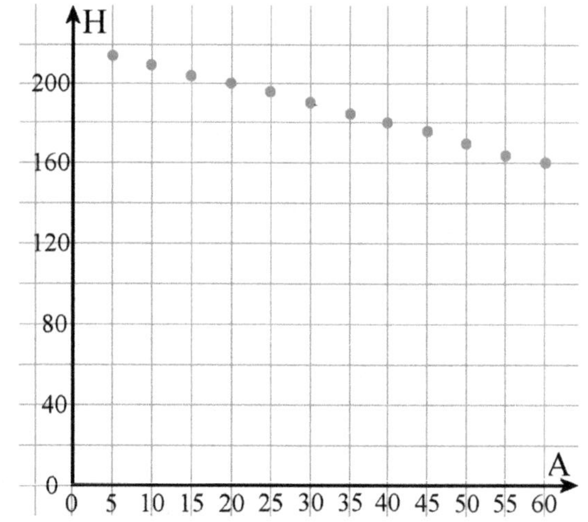

8. a.

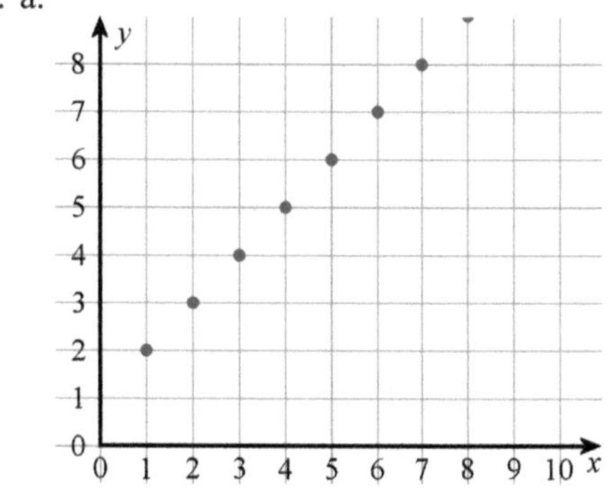

x	1	2	3	4	5	6	7	8	9
y	2	3	4	5	6	7	8	9	10

Equation: $y = x + 1$

b.

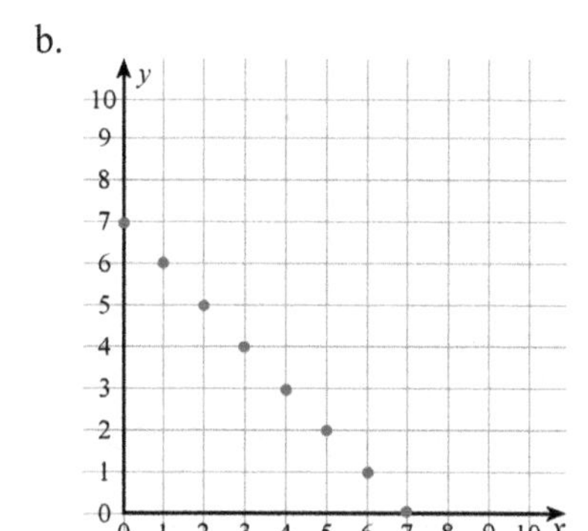

x	0	1	2	3	4	5	6	7
y	7	6	5	4	3	2	1	0

Equation: $y = 7 - x$

Using Two Variables, cont.

Puzzle corner:
a. $y = 10 - (x/2)$ or $y = 10 - (1/2)x$ or $y = -(1/2)x + 10$
b. $y = 2x + 1$

x	0	1	2	3	4	5	6	7	8	9	10
y	10	9.5	9	8.5	8	7.5	7	6.5	6	5.5	5

x	0	1	2	3	4	5
y	1	3	5	7	9	11

Mixed Review, p. 85

1. a. 100,000
   b. 81
   c. 400,000

2. 10^7 is ten million.

3. a. $A = 3 \text{ km} \cdot 3 \text{ km} = 9 \text{ km}^2$
   b. $V = 2 \text{ in} \cdot 2 \text{ in} \cdot 2 \text{ in} = 8 \text{ in}^3$

4. a. 4×9 cm = 36 cm. The perimeter is 36 cm.
   b. $48 \text{ m} \div 4 = 12$ m. The area is $12 \text{ m} \times 12 \text{ m} = 144 \text{ m}^2$.

5. a. 5,051,000,027,000
   b. 21,650,099,000,056

6. 6,002,001

7. a. $5 \cdot 10^4 + 4 \cdot 10^3$
   b. $2 \cdot 10^6 + 9 \cdot 10^4 + 3 \cdot 10^1$

8. a. $5s \div 8$
   b. $7(x + 8)$
   c. $8 - y$
   d. $(x - 8)^2$

9. Estimations may vary.

a. 591 · 57,200	b. 435,212 + 9,319,290
Estimation: 600 · 57,000 = 34,200,000 OR 600 · 60,000 = 36,000,000 Exact: 33,805,200 Error of estimation: 394,800 OR 2,194,800	Estimation: 435,000 + 9,320,000 = 9,755,000 Exact: 9,754,502 Error of estimation: 498

10.

a. $4 \cdot 50 + \dfrac{310}{2} = 355$	b. $\dfrac{4{,}800}{60} - (70 - 20) = 30$

11. The price is reduced by 2/10. So \$120 represents the remaining 8/10 of the original cost.
    We can find 1/10 of the price by dividing the remaining price by 8: \$120 ÷ 8 = \$15.
    Now, \$15 is 1/10 of the original cost; therefore the original price was 10 · \$15 = \$150.

12. 7123 R73

Chapter 2 Review, p. 87

1. a. $(6 - x)^2$

   b. $\dfrac{5}{x+6}$

   c. $3(5 - p)$

2. a. 113 b. 200
   c. 9 d. 560

3. a. 28 b. 91

4. a. $p \div 3$

   b. $\$3 + 6c$ *or* $6c + \$3$

5.

$2x + 17$	$8 = 8$	$y < 5$	$4x - 3 = 8$	$\frac{4}{5}x - 16$	$4x + y^2 \geq 9$	$M = \dfrac{44 - x}{5}$
expression	equation	inequality	equation	expression	inequality	equation

6.

a. $3t + 3$	b. $5d$
c. x^3	d. $6x - 6$
e. $16z^3$	f. $14x^2 + 5$

7.

a. $A = 9s^2$ $P = 12s$	b. $A = 6x^2$ $P = 10x$

8. a. $6x + 21$
   b. $72b + 40$

9.

a. $5x + 10 = 5(x + 2)$	b. $6y + 10 = 2(3y + 5)$
c. $24b + 4 = 4(6b + 1)$	d. $25w + 40 = 5(5w + 8)$

10. a. 112
    b. 116
    c. 72
    d. 60 4/13
    e. 328
    f. 70

11. a. $25q = 1675$. $q = 67$. There are 67 quarters.

    b. $2(x + 21) = 128$ OR $x + x + 21 + 21 = 128$ OR $2x + 21 + 21 = 128$. $x = 43$. The other side is 43 meters long.

12. The temperature is 77 degrees Fahrenheit.

13. a. $x < 57$
    b. $x \geq 30$

14. a. 23, 30, 55, 44

    b. 2, 4, 6

15. $y = x + 3$

x	1	2	3	4	5	6
y	4	5	6	7	8	9

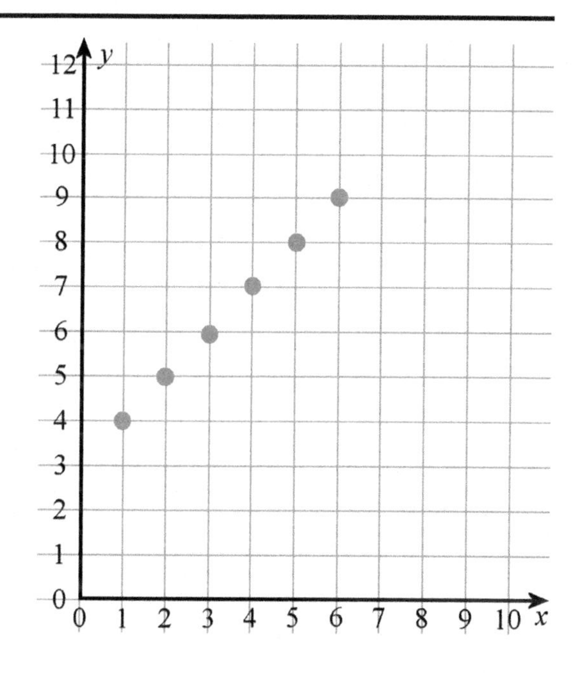

16. $y = x + 3$

t (hours)	0	1	2	3	4	5	6
d (miles)	0	70	140	210	280	350	420

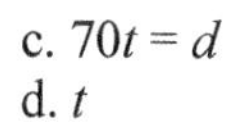

c. $70t = d$
d. t

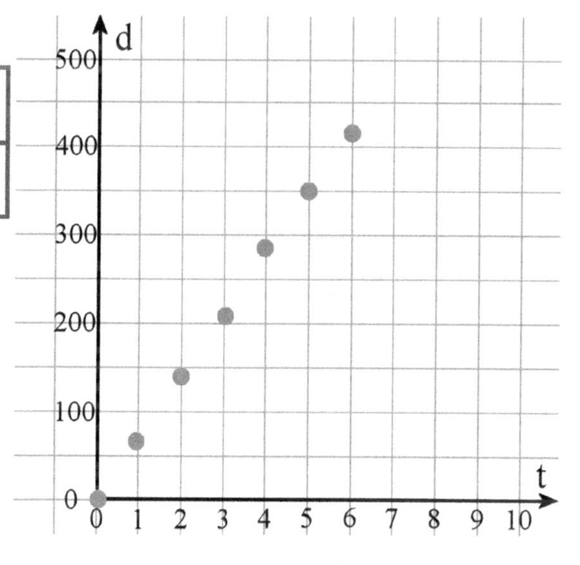

Chapter 3: Decimals

Place Value with Decimals, p. 95

1.

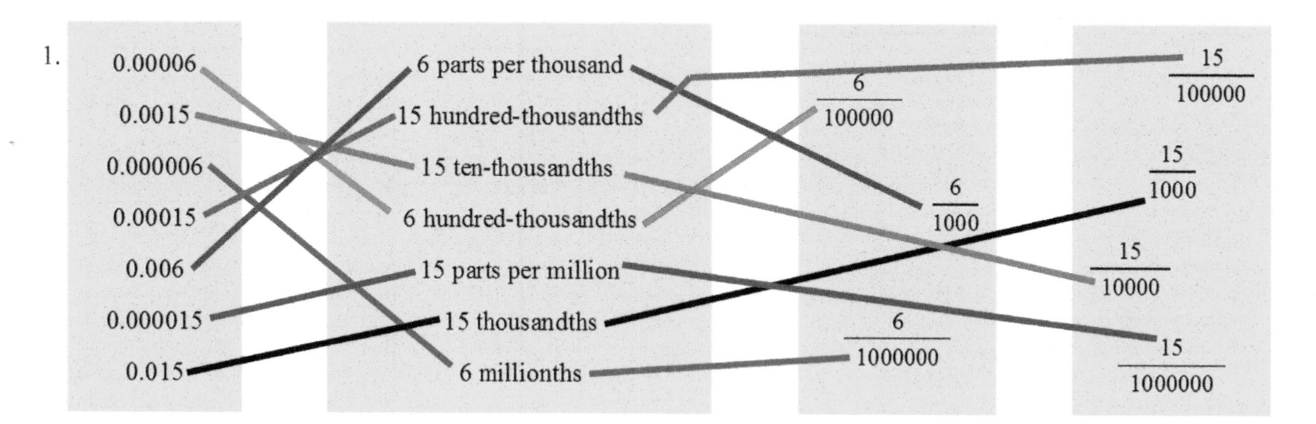

2. a. 0.003 b. 3.4
   c. 1.001934 d. 0.0034
   e. 0.000907 f. 0.00837
   g. 0.52 h. 0.00008
   i. 3.017 j. 0.000091
   k. 1.056 l. 2.028319
   m. 0.0291 n. 4.000005

3.a. 9/100 b. 5/1000 c. 45/1000
   d. 371/100,000 e. 2381/100,000 f. 30,078/10,000
   g. 29,302/10,000 h. 2,003,814/1,000,000 i. 53,925,012/10,000,000
   j. 31/10,000,000 k. 3,294,819/1,000,000 l. 4,500,032/100,000

4. a. $2.67 = 2 \times 1 + 6 \times \frac{1}{10} + 7 \times \frac{1}{100}$

   b. $0.594 = 5 \times \frac{1}{10} + 9 \times \frac{1}{100} + 4 \times \frac{1}{1{,}000}$

   c. $45.6 = 4 \times 10 + 5 \times 1 + 6 \times \frac{1}{10}$

   d. $0.004923 = 4 \times \frac{1}{1{,}000} + 9 \times \frac{1}{10{,}000} + 2 \times \frac{1}{100{,}000} + 3 \times \frac{1}{1{,}000{,}000}$

   e. $0.00000506 = 5 \times \frac{1}{1{,}000{,}000} + 6 \times \frac{1}{100{,}000{,}000}$

5. a. 65.286 b. 5.055009
   c. 700.07103 d. 0.010304
   e. 6.5903 f. 2.02101

Comparing Decimals p. 97

1. a. $0.067 > 0.0098$ b. $0.0005 < 0.005$ c. $1.828 > 1.0828$
   d. $2.504040 < 2.505404$ e. $8.00014 < 8.004$ f. $0.91701 > 0.917005$

Comparing Decimals, cont.

2.

a. 0.05	0.009	0.1	b. 1.04	1.2013	1.1	c. 0.905	0.86948	0.9
d. 0.0004	0.0000337		e. 9.082	9.1	9.09	f. 0.288391	0.284857	

3. a. 0.0087 > 0.0009   b. 1.005830 > 1.002301   c. 1.270038 < 1.270110
d. 0.0000020 > 0.0000004   e. 0.000026 > 0.000010   f. 0.450 > 0.300

4. a.

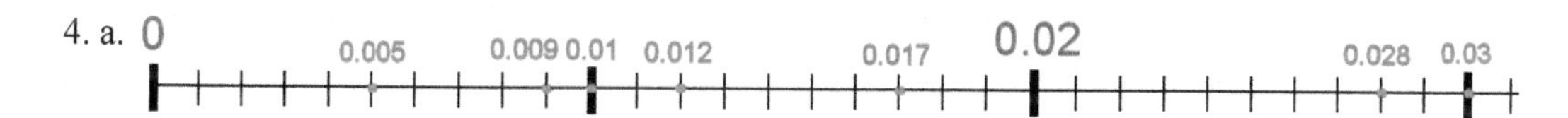

Note: The student is expected to *reason* what the midpoints would be, based on the visual number line, and is not required to *calculate* them. The calculations below are included as enrichment to show how it *could* be done and how it gives exactly the same results.

b. The midpoint of 0 and 0.01 is exactly (0 + 0.01) / 2 = 0.005
c. The midpoint of 0.01 and 0.02 is exactly (0.01 + 0.02) / 2 = 0.015
d. The midpoint of 0 and 0.001 is exactly (0 + 0.001) / 2 = 0.0005
e. The midpoint of 0.001 and 0.002 is exactly (0.001 + 0.002) / 2 = 0.0015

5.

a. 0.0491 < 0.2930 < 0.3100	b. 1.043 < 1.304 < 1.340
c. 3.00028 < 3.00890 < 3.00980	d. 0.000003 < 0.000020 < 0.000023
e. 1.0980 < 1.0987 < 1.1070	f. 0.0400 < 0.0450 < 0.0456

6. a. 0.400 + 0.008 = 0.408
b. 0.20 + 0.07 = 0.27
c. 4.001 + 0.007 = 4.008
d. 0.005 + 0.060 = 0.065

7.

Number	0.4	0.35	0.027	1.297	5.99	0.606
+ 0.1	0.5	0.45	0.127	1.397	6.09	0.706
+ 0.01	0.41	0.36	0.037	1.307	6	0.616
+ 0.001	0.401	0.351	0.028	1.298	5.991	0.607

Add and Subtract Decimals, p. 99

1.

a.

O		t	h	th	t-th
0	.	0	0	2	

1 tenth more: 0.002 + 0.100 = 0.102
1 thousandth less: 0.002 − 0.001 = 0.001
1 ten-thousandth more: 0.0020 + 0.0001 = 0.0021

b.

O		t	h	th	t-th	h-th
0	.	8	5			

2 hundredths less: 0.85 − 0.02 = 0.83
2 ten-thousandths more: 0.8500 + 0.0002 = 0.8502
2 hundred-thousandths more: 0.85000 + 0.00002 = 0.85002

2.

a. 0.2 + 0.8 = 1.0 b. 0.20 + 0.08 = 0.28 c. 0.2000 + 0.0008 = 0.2008	d. 0.03 + 0.06 = 0.09 e. 0.0300 + 0.0006 = 0.0306 f. 0.03000 + 0.00006 = 0.03006	g. 0.090 + 0.007 = 0.097 h. 0.900 + 0.007 = 0.907 i. 0.00009 + 0.00700 = 0.00709

Add and Subtract Decimals, cont.

3.

a. 1.30 + 0.56 = <u>1.86</u>	b. 0.20 + 0.27 = <u>0.47</u>	c. 3.19 + 0.50 = <u>3.69</u>	d. 2.289 − 0.100 = <u>2.189</u>

4. a. 0.25, 0.28, 0.31, <u>0.34</u>, <u>0.37</u>, <u>0.40</u>, <u>0.43</u>, <u>0.46</u>, <u>0.49</u>
   b. 3.275, 3.28, 3.285, <u>3.29</u>, <u>3.295</u>, <u>3.3</u>, <u>3.305</u>, <u>3.31</u>, <u>3.315</u>

5. a. 0.15 + 0.20 = ~~0.17~~ <u>0.35</u> b. 1.06 + 0.04 = <u>1.10</u> was correct. c. 0.90 − 0.08 = ~~0.10~~ <u>0.82</u>

6. a. 0.5 − 0.2 = <u>0.3</u>	b. 0.5 − 0.02 = <u>0.48</u>	c. 0.5 − 0.002 = <u>0.498</u>

7. As seen below, other tricks you can use when adding or subtracting in columns are:
   - line the decimal points and then fill the missing spaces with zeroes
   - group the decimal digits in threes, to help keep track of all the digits

7. a.
```
   6.907 00
 − 4.800 56
 ----------
   2.106 44
```

b.
```
   2.000 000
 + 9.082 000
 + 0.038 284
 + 4.502 800
 -----------
  15.623 084
```

c.
```
  410.000 00
 − 25.600 00
 −  4.593 84
 -----------
  379.806 16
```

8. a. 0.200 + 0.035 = <u>0.235</u> b. 0.0004 + 0.0400 = <u>0.0404</u>
   c. 0.70000 + 0.00205 − 0.01800 = <u>0.68405</u> d. 9.0000 + 0.0009 − 0.5000 = <u>8.5009</u>

9. a. x = 2.3000 − 0.0095 = <u>2.2905</u>
   b. x = 0.00311 + 0.39192 = <u>0.39503</u>
   c. x = 2.00480 − 1.28029 = <u>0.72451</u>

Rounding Decimals, p. 101

1. a. 2.7 b. 3.055 c. 2.27 d. 0.049

2. a. 7.25 b. 0.02 c. 1.36 d. 4.98

3. 7.249 b. 0.027 c. 1.395 d. 4.908

4. a. \$2 + \$1 + \$7 + \$12 = \$22
   b. \$30 − \$22 = \$8
   c. Exact cost: \$1.82 + \$0.89 + \$6.72 + 2 × \$6.12 = \$21.67. Error of estimation: \$0.33.

5. a. 1.0 b. 2.060 c. 6.20
   d. 0.050 e. 7.240 f. 0.200
   g. 4.00 h. 0.100 i. 0.0090

6. a. 0.3840 b. 2.39 c. 0.03900
   d. 938 e. 710 f. 1000

7.

Number:	0.289940	1.293854	2.5949405	0.394040	2.299775
...three decimals	0.290	1.294	2.595	0.394	2.300
...four decimals	0.2899	1.2939	2.5949	0.3940	2.2998
...five decimals	0.28994	1.29385	2.59494	0.39404	2.29978

Rounding Decimals, cont.

8.

a. 0.1539204 + 0.23609

Estimate: 0.15 + 0.24 = 0.39

Exact:

$$\begin{array}{r} 0.1539204 \\ +0.23609 \\ \hline 0.3900104 \end{array}$$

b. 1.39821 + 0.2831

Estimate: 1.40 + 0.28 = 1.68

Exact:

$$\begin{array}{r} 1.39821 \\ +0.2831 \\ \hline 1.68131 \end{array}$$

c. 3.4822 + 3 − 4.5078

Estimate:
3.48 + 3 − 4.51 = 6.48 − 4.51 = 1.97

Exact:

$$\begin{array}{r} \scriptstyle 11 \\ \scriptstyle 5\ 14\ 7\ \cancel{1}12 \\ \cancel{6}.\cancel{4}\,\cancel{8}\,\cancel{2}\,\cancel{2} \\ -4.5078 \\ \hline 1.9744 \end{array}$$

d. 2.917328 − 0.302849 − 1.0549

Estimate:
2.92 − 0.30 − 1.05 = 2.62 − 1.05 = 1.57

Exact:

$$\begin{array}{r} 0.302849 \\ +1.0549 \\ \hline 1.357749 \end{array} \qquad \begin{array}{r} \scriptstyle 10\ 16\ 12\ 11 \\ \scriptstyle 8\ \cancel{0}\ \cancel{6}\ \cancel{2}\ \cancel{1}18 \\ 2.\cancel{9}\,\cancel{1}\,\cancel{7}\,\cancel{3}\,\cancel{2}\,\cancel{8} \\ -1.357749 \\ \hline 1.559579 \end{array}$$

Mystery Number: 5.448

Review: Multiply and Divide Decimals Mentally, p. 104

1. a. 7 × 0.6 = 4.2
   7 × 0.06 = 0.42
   7 × 0.006 = 0.042

   b. 4 × 1.5 = 6.0
   4 × 0.15 = 0.60
   4 × 0.0015 = 0.0060

   c. 3 × 0.05 = 0.15
   3 × 0.005 = 0.015
   3 × 0.0005 = 0.0015

   d. 9 × 0.8 = 7.2
   9 × 0.008 = 0.072
   9 × 0.00008 = 0.00072

2.

a. 0.36 ÷ 4 = 0.09 0.09 × 4 = 0.36	b. 3.5 ÷ 7 = 0.5 0.5 × 7 = 3.5	c. 0.008 ÷ 2 = 0.004 0.004 × 2 = 0.008	d. 0.099 ÷ 9 = 0.011 0.011 × 9 = 0.099
e. 0.0046 ÷ 2 = 0.0023 0.0023 × 2 = 0.0046	f. 0.0024 ÷ 3 = 0.0008 0.0008 × 3 = 0.0024	g. 0.00049 ÷ 7 = 0.00007 0.00007 × 7 = 0.00049	h. 0.144 ÷ 12 = 0.012 0.012 × 12 = 0.144

3.

a. 0.36 ÷ 0.06 = 6 6 × 0.06 = 0.36	b. 1.8 ÷ 0.2 = 9 9 × 0.2 = 1.8	c. 0.054 ÷ 0.006 = 9 9 × 0.006 = 0.054

4. a. 4 ÷ 0.5 = 8 b. 2 ÷ 0.4 = 5 c. 0.56 ÷ 0.07 = 8
   d. 0.0012 ÷ 0.0004 = 3 e. 0.015 ÷ 0.005 = 3 f. 0.0032 ÷ 0.0008 = 4

5.

a. 0.30 ÷ 0.15 = 2 2 × 0.15 = 0.30	b. 0.040 ÷ 0.005 = 8 8 × 0.005 = 0.040	c. 0.40 ÷ 0.02 = 20 20 × 0.02 = 0.40
d. 0.600 ÷ 0.002 = 300 300 × 0.002 = 0.600	e. 0.70 ÷ 0.01 = 70 70 × 0.01 = 0.70	f. 0.00800 ÷ 0.00002 = 400 400 × 0.00002 = 0.00800

6. a. 15.3 m ÷ 0.3 m = 51 pieces.
   b. $63.96 ÷ 3 = $21.32. Divide in parts: ($60 ÷ 3) + ($3 ÷ 3) + ($0.90 ÷ 3) + ($0.06 ÷ 3) = $21.32.

7. a. 2.5 ÷ 5 = 0.5 b. 3.0 ÷ 0.6 = 5 c. 1.02 ÷ 2 = 0.51
   d. 0.048 ÷ 0.008 = 6 e. 0.60 ÷ 0.02 = 30 f. 0.0056 ÷ 7 = 0.0008
   g. 4.018 ÷ 2 = 2.009 h. 0.0306 ÷ 3 = 0.0102 i. 0.5055 ÷ 5 = 0.1011

Review: Multiply Decimals by Decimals, p. 106

1.

a. $0.2 \times 0.8 = 0.16$	b. $0.7 \times 0.12 = 0.084$	c. $0.03 \times 0.9 = 0.027$
d. $0.004 \times 0.5 = 0.0020$	e. $0.02 \times 0.0009 = 0.000018$	f. $0.011 \times 0.06 = 0.00066$

2. a. $1.1 \times 0.02 \times 0.5 = 0.0110$
   b. $0.3 \times 0.07 \times 0.2 = 0.0042$
   c. $5 \times 0.02 \times 0.004 = 0.00040$
   d. $0.3 \times 4 \times 0.002 = 0.0024$
   e. $0.3 \times 0.3 \times 0.3 = 0.027$
   f. $4 \times 0.4 \times 0.0005 = 0.00080$

3. a. 0.3 m × \$3.20/m = \$0.96. Multiply 3 × 32 = 96. The only reasonable answer is \$0.96 (\$9.60 would not make sense).
   b. 1.2 kg × \$50/kg = \$60. Multiply 12 × 5 = 60. The only reasonable answer is \$60 - slightly more than \$50, since 1.2 kg is slightly more than 1 kg.

4.

a. $3k = 3 \times 0.008 = 0.024$	b. $s^2 = 0.3^2 = 0.09$	c. $k^2 = 0.008^2 = 0.000064$
d. $s - k = 0.3 - 0.008 = 0.292$	e. $sk + 1 = 0.008 \times 0.3 + 1 = 1.0024$	f. $2s + 2k = 2 \times 0.3 + 2 \times 0.008 = 0.616$

5.

a. $0.4 \times 0.03 = 0.012$ ↓ $\frac{4}{10} \times \frac{3}{100} = \frac{12}{1000}$	b. $0.009 \times 0.3 = 0.0027$ ↓ $\frac{9}{1000} \times \frac{3}{10} = \frac{27}{10000}$	c. $0.011 \times 0.0006 = 0.0000066$ ↓ $\frac{11}{1000} \times \frac{6}{10{,}000} = \frac{66}{10{,}000{,}000}$
d. $1.1 \times 0.002 = 0.0022$ ↓ $\frac{11}{10} \times \frac{2}{1000} = \frac{22}{10{,}000}$	e. $0.21 \times 0.0004 = 0.000084$ ↓ $\frac{21}{100} \times \frac{4}{10{,}000} = \frac{84}{1{,}000{,}000}$	f. $0.005 \times 0.005 = 0.000025$ ↓ $\frac{5}{1000} \times \frac{5}{1000} = \frac{25}{1{,}000{,}000}$

6. 800 × 1.4 = 1120 and 1000 × 1.4 = 1400. The new image would be 1120 by 1400 pixels.

7.

a. $y = 0.006 \div 3 = 0.002$	b. $z = 0.00033 \div 3 = 0.00011$	c. $d = 0.72 \div 0.9 = 0.8$
d. $x = 2 \times 0.008 = 0.016$	e. $z = 0.5 \times 0.07 = 0.035$	f. $h = 2.1 \times 0.001 = 0.0021$

8. Please check the student's work. There can be various answers depending on how they round the numbers.

a. 0.37×0.91 $\approx 0.4 \times 0.9 = 0.36$	b. 1.205×0.51 $\approx 1.2 \times 0.5 = 0.6$	c. 3.93×0.043 $\approx 4 \times 0.04 = 0.16$
$0.37 \times 0.91 = 0.3367$	$1.205 \times 0.51 = 0.61455$	$3.93 \times 0.043 = 0.16899$

9. a. $12.345 \times 678.9 = 8381.0205$ b. $123.45 \times 0.6789 = 83.810205$ c. $1.2345 \times 67.89 = 83.810205$

10. a. $3.4 \times 0.28 =$ 0.952. Estimate: $3 \times 0.3 =$ 0.9
    b. $0.455 \times 6.4 =$ 2.912. Estimate: $0.5 \times 6 =$ 3
    c. $3.08 \times 0.0034 =$ 0.010472. Estimate: $3 \times 0.003 =$ 0.009
    d. $0.007 \times 0.000135 =$ 0.000000945. Estimate: $0.007 \times 0.0001 =$ 0.0000007

11. a. $x = 4.5 \times 0.6 = 2.7$
    b. $z = 2.2 \times 0.002 = 0.0044$

Review: Long Division with Decimals, p. 109

1. a. 1.57 Check: $5 \times 1.57 = 7.85$
   b. 0.687 Check: $3 \times 0.687 = 2.061$
   c. 0.87957 Check: $0.87957 \times 7 = 6.15699 \approx 6.157$

2. a. $7.1 \div 6 \approx 1.183$ b. $1.3 \div 3 \approx 0.433$ c. $2.509 \div 7 \approx 0.358$

3. a. $7 \div 11 \approx 0.636$ b. $21 \div 8 = 2.625$ c. $14.2 \div 5 = 2.84$
   d. $45.08 \div 9 \approx 5.009$ e. $48.44 \div 14 = 3.46$ f. $52.7 \div 23 \approx 2.291$

4. a. $4.1314 \div 7 = 0.5902$ b. $46.08 \div 9 = 5.12$
   c. $0.342 \div 6 = 0.057$ d. $125 \div 13 \approx 9.615$
   e. $212.5 \div 23 \approx 9.239$ f. $460 \div 51 \approx 9.020$

Problem Solving with Decimals, p. 111

1. Two fathoms are 2×6 feet = 12 feet, and 3 2/3 m are about 3.667 m or 3.667×3.28 feet $\approx$ 12.03 feet. Since 12.03 feet > 12 feet, 3 2/3 m are longer than two fathoms (by about a third of an inch).

2. Jack, John, and Jerry each got $\$200.00 \div 3 = \$66.66$ (with two cents left over).

3. 45 books. Five books would weigh 2 kg. You can pack nine times that many books, or 45 books, into a 18-kg suitcase.

4. The average is the sum of all of the student's scores divided by the total number of students:
   $(21 + 15 + 18 + 29 + 19 + 34 + 39 + 21 + 11 + 8 + 15 + 28 + 15 + 11 + 12) / 15 \approx 19.73$.

5. $A = 2.4\text{ m} \times 1.2\text{ m} + 1.2\text{ m} \times 2.4 + 1.8 \times 1.8 = 9\text{ m}^2$. The area is 9 square meters.
   $P = 2.4\text{ m} + 1.2\text{ m} + 1.2\text{ m} + 2.4\text{ m} + 0.6\text{ m} + 1.8\text{ m} + 1.8\text{ m} + 5.4\text{ m} = 16.8\text{ m}$. The perimeter is 16.8 meters.

6. One 40-kg crate will hold the weight of (40 kg) $\div$ (1.2 kg / blender) = 33 1/3 blenders. Since they will pack only whole blenders, they can pack 33 blenders into each shipping crate.

7.

Item and price	Unit price	What would this cost...?	
32 oz of orange juice for $3.65	$0.11 per oz	12 oz of orange juice	$1.32
3 lb of chicken for $3.24	$1.08 per lb	5.7 lb of chicken	$6.16
4 lb of bananas for $1.99	$0.50 per lb	2.5 lb of bananas	$1.25

Fractions and Decimals, p. 113

1. a. 9/100 b. 5/1000 c. 45/1000
   d. 371/100,000 e. 2381/100,000 f. 31/10,000,000

2.

a. $\frac{29,302}{10,000} = 2\frac{9302}{10,000}$	b. $\frac{2,003,814}{1,000,000} = 2\frac{3814}{1,000,000}$
c. $\frac{53,925,012}{10,000,000} = 5\frac{3925012}{10,000,000}$	d. $\frac{30,078}{10,000} = 3\frac{78}{10,000}$
e. $\frac{3,294,819}{1,000,000} = 3\frac{294819}{1,000,000}$	f. $\frac{4,500,032}{100,000} = 45\frac{32}{100,000}$

3. a. 0.36 b. 5.009 c. 1.045
   d. 0.3908 e. 2.00593 f. 0.005903
   g. 45.039034 h. 43.5112 i. 4.50683

4.

a. × 2: $\frac{1}{5} = \frac{2}{10} = 0.2$ (× 2)	b. × 125: $\frac{1}{8} = \frac{125}{1000} = 0.125$ (× 125)	c. × 5: $1\frac{1}{20} = 1\frac{5}{100} = 1.05$ (× 5)
d. × 4: $3\frac{9}{25} = 3\frac{36}{100} = 3.36$ (× 4)	e. ÷ 2: $\frac{12}{200} = \frac{6}{100} = 0.06$ (÷ 2)	f. × 25: $8\frac{3}{4} = 8\frac{75}{100} = 8.75$ (× 25)
g. $4\frac{3}{5} = 4\frac{6}{10} = 4.6$	h. $\frac{13}{20} = \frac{65}{100} = 0.65$	i. $\frac{7}{8} = \frac{875}{1000} = 0.875$
j. $\frac{11}{125} = \frac{88}{1000} = 0.088$	k. $\frac{24}{400} = \frac{6}{100} = 0.06$	l. $\frac{95}{500} = \frac{19}{100} = 0.19$

5. a. 9/20 or 0.45 b. 1 27/50 or 1.54 c. 3 9/10 or 3.9 d. 107/200 or 0.535
   e. 77/100 or 0.77 f. 5 1/40 or 5.025 g. 47/100 or 0.47 h. 79/100 or 0.79

6. a. 2 ÷ 9 = 0.222 b. 3 ÷ 7 = 0.429 c. 7 ÷ 16 = 0.438

7. a. 1 ÷ 11 = 0.091 b. 3 ÷ 23 = 0.130 c. 47 ÷ 56 = 0.839

8.
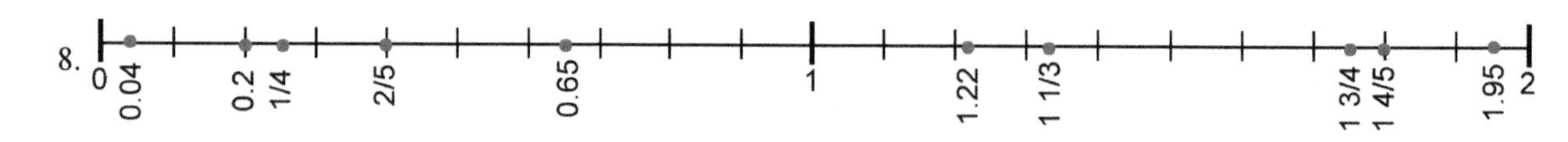

9. 900 g + 750 g = 1650 g or 1.65 kg.

10. a. 14.375 inches by 20.375 inches. b. The puzzle is 292.890625 in^2.

11. \$11.45 × 1.75 = 20.0375 The flax seed cost \$20.04.

12. Method one: Divide \$12.95 by eight, then multiply the answer by three.
    Method two: Three eighths is 0.375. Multiply 12.95 by 0.375. You will need to round your answer to two decimals to show dollars and cents.

13. a. $^3/_{16}$ or 0.1875
    b. 1 $^1/_8$ or 1.125

Multiply and Divide by Powers of Ten p. 116

1.

a. $10 \times 3.84 = 38.4$ $1{,}000 \times 3.84 = 3{,}840$	b. $100 \times 0.09 = 9.$ $10{,}000 \times 0.594 = 5{,}940$	c. $10{,}000 \times 3.84 = 38{,}400$ $1{,}000 \times 0.0038 = 3.8$
d. $10^3 \times 1.09 = 1{,}090$ $10^4 \times 1.09 = 10{,}900$	e. $10^3 \times 0.0075 = 7.5$ $10^5 \times 0.0075 = 750$	f. $10^7 \times 0.0021 = 21{,}000$ $10^6 \times 4.8 = 4{,}800{,}000$

2.

a. $1.5 \div 10 = 0.15$ $0.43 \div 10 = 0.043$	b. $1.08 \div 100 = 0.0108$ $2.3 \div 100 = 0.023$	c. $56 \div 10 = 5.6$ $56 \div 1000 = 0.056$
d. $0.69 \div 10^3 = 0.00069$ $51.0 \div 10^4 = 0.0051$	e. $2.9 \div 10^5 = 0.000029$ $4{,}500 \div 10^6 = 0.0045$	f. $67.8 \div 100 = 0.678$ $251 \div 10^4 = 0.0251$

3.

a. $0.23 \div 100 = 0.0023$ $1{,}400 \div 10{,}000 = 0.14$ $3.892 \div 1000 = 0.003892$	b. $1{,}000 \times 97.201 = 97{,}201$ $10^6 \times 0.004835 = 4{,}835$ $10^4 \times 3.49284 = 34{,}928.4$	c. $10^6 \times 34.2958 = 34{,}295{,}800$ $10^5 \times 0.00293 = 293$ $10^7 \times 2.19304 = 21{,}930{,}400$

4.

a. $0.15 \times 100 = 15$ $0.932 \times 10 = 9.32$	b. $100 \times 30.4 = 3{,}040$ $1{,}000 \times 5.5 = 5{,}500$
c. $0.029 \times 100{,}000 = 2{,}900$ $0.0006 \times 1{,}000{,}000 = 600$	d. $100{,}000 \times 0.34 = 34{,}000$ $100 \times 0.00478 = 0.478$

e. $\frac{17}{1{,}000} = 0.017$	f. $\frac{2.3}{100} = 0.023$	g. $\frac{412}{100} = 4.12$	h. $\frac{0.58}{10} = 0.058$

Review: Divide Decimals by Decimals, p. 118

1. The answers are the same. In each case, moving the decimal point of *both* the dividend and the divisor the same number of places makes no change in the quotient.

a. $120 \div 20 = 6$ b. $12 \div 2 = 6$ c. $1.2 \div 0.2 = 6$ d. $0.12 \div 0.02 = 6$	e. $28 \div 4 = 7$ f. $2.8 \div 0.4 = 7$ g. $0.28 \div 0.04 = 7$ h. $0.028 \div 0.004 = 7$

2.

a. $0.8 \div 0.02$ $8 \div 0.2$ $80 \div 2 = 40$	b. $12 \div 0.4$ $120 \div 4 = 30$	c. $4.5 \div 0.05$ $45 \div 0.5$ $450 \div 5 = 90$

3.

a. $1.6 \div 0.04$ $160 \div 4 = 40$	b. $2.6 \div 0.2$ $26 \div 2 = 13$	c. $36 \div 0.009$ $36000 \div 9 = 4{,}000$
d. $0.6 \div 0.003$ $600 \div 3 = 200$	e. $5.4 \div 0.009$ $5400 \div 9 = 600$	f. $0.5 \div 0.005$ $500 \div 5 = 100$

Divide Decimals by Decimals, cont.

4\.

a.	b.	c.
$\frac{9}{0.3} = \frac{90}{3} = 30$ (× 10 top and bottom)	$\frac{2}{0.05} = \frac{200}{5} = 40$ (× 100 top and bottom)	$\frac{0.3}{0.006} = \frac{300}{6} = 50$ (× 1000 top and bottom)

5\.

a.	b.	c.
$\frac{3.4}{0.002} = \frac{3400}{2} = 1700$ (× 1000 top and bottom)	$\frac{0.56}{0.0008} = \frac{5600}{8} = 700$ (× 10000 top and bottom)	$\frac{0.15}{0.0003} = \frac{1500}{3} = 500$ (× 10000 top and bottom)

6\. The following are all equal to 0.2: a. $\frac{2}{10}$ c. $\frac{20}{100}$ e. $\frac{200}{1000}$

7\. Answers will vary. Please check the student's work. You can build the divisions by thinking of multiplication: 5 times some decimal number equals something. For example:

$2.5 \div 0.5 = 5$ $0.01 \div 0.002 = 5$ $7.5 \div 1.5 = 5$

8\. a. $x = 2.5$ b. $y = 70$ c. $z = 500$

9\. a. 92 b. 34.8 c. 6.26 d. 133

Divide Decimals by Decimals 2, p. 121

1\. a. $5,720 \div 4 = 1,430$ b. $23,880 \div 6 = 3,980$

2\. a. $x = 14.7$
b. $x = 459.2$

3\. a. 841.05 b. 2770.31
c. 1960 d. 2652.78

4\. ($5.56 × 2.7) ÷ 3 = $5.00 with one cent left over.
Two of the three people would pay $5.00 and the other person would pay $5.01.

5\. $\frac{2}{0.8} = \frac{20}{8} = 2\ 4/8 = 2\ 1/2 = 2.5$

6\. a. np

b. $\frac{np}{10}$

c. $\frac{np}{10} \cdot 9$ or $\frac{9np}{10}$ or $0.9np$ or $np - np/10$ or $np - 0.1np$

Puzzle corner: Answers will vary. Examples:

a.	b.	c.	d.
$\frac{732}{5} = \frac{1464}{10} = 146.4$ (× 2 top and bottom)	$\frac{842}{50} = \frac{1684}{100} = 16.84$ (× 2 top and bottom)	$\frac{6050}{25} = \frac{24,200}{100} = 242$ (× 4 top and bottom)	$\frac{250}{4} = \frac{125}{2} = 62.5$ (÷ 2 top and bottom)

Convert Customary Measuring Units, p. 123

1. a. the upper b. the lower c. the lower d. the upper

2.

a. 564 ft = 0.11 mi b. 45,000 ft = 8.52 mi	c. 3,400 yd = 1.93 mi d. 7.8 mi = 41,184 ft	e. 0.28 mi = 1478.4 ft f. 10.17 mi = 17899.2 yd

3.

a. 3 in = 0.25 ft b. 21 in = 1.75 ft	c. 14.7 ft = 176.4 in d. 0.8 ft = 9.6 in	e. 281 in = 23.42 ft f. 7 1/3 ft = 88 in

4.

a. 5 oz = 0.31 lb b. 35 oz = 2.19 lb	c. 3.6 lb = 57.6 oz d. 0.391 lb = 6.26 oz	e. 127 oz = 7.94 lb f. 6 3/4 lb = 108 oz

5.

a. 6.4 gal = 25.6 qt b. 78 fl. oz. = 2.44 qt c. 2.3 qt = 73.6 fl. oz.	d. 0.56 qt = 17.92 fl. oz. e. 560 qt = 140 gal f. 3.2 T = 6,400 lb	g. 0.054 T = 108 lb h. 1,200 lb = 0.6 T i. 6,750 lb = 3.38 T

6.

a. 2 ft 6 in = 30 in b. 7 ft 11 in = 95 in	c. 162 in = 13 ft 6 in d. 79 in = 6 ft 7 in	e. 254 in = 21 ft 2 in f. 1,028 in = 85 ft 8 in

7.

a. 6 lb 9 oz = 105 oz b. 11 lb 12 oz = 188 oz	c. 86 oz = 5 lb 6 oz d. 145 oz = 9 lb 1 oz	e. 483 oz = 30 lb 3 oz f. 591 oz = 36 lb 15 oz

8.

a. 2.7 ft = 2 ft 8 in b. 10.2 ft = 10 ft 2 in	c. 3.15 ft = 3 ft 2 in d. 7.8 lb = 7 lb 13 oz	e. 55.46 lb = 55 lb 7 oz f. 8.204 lb = 8 lb 3 oz

9. a. 13 lb 11 oz b. 109 ft 2 in c. 14 qt 30 oz d. 45 ft 6 in e. 12 lb 3 oz f. 2 h 49 min

10. 4 ft 3 in − (3 × 2 $^3/_8$) in = 4 ft 3 in − 7 1/8 in = 3 ft 7 $^7/_8$ in or about 3 ft 8 in.
Jack was 3 feet 8 inches tall three years ago.

11. The room is 40 ft × 12 in/ft = 480 inches wide. Since each chair is 21 inches, and 480 ÷ 21 ≈ 22.857, you can place 22 chairs in one row.

The two 3-ft aisles take up 6 ft × 12 in/ft = 72 inches, so there are 480 in − 72 in = 408 inches left for the chairs. This time, since 408 ÷ 21 ≈ 19.429, you can fit 19 chairs in one row.

12. 4 × 32 ÷ 6 = 21.33. Jack would need 21 glasses for the tea.

13. (60 × 16) ÷ (2 × 16 + 3) = 27.43. You can pack 27 math books in the box.

14. a. $5.69 ÷ 13 = $0.437692307... per ounce $13.99 ÷ 32 = $0.4371875 per ounce
The 1-quart bottle is the better deal.
b. $0.35 ÷ 12 = $0.0291666... per ounce. $4.10 ÷ (8 × 16) = $0.03203125 per ounce.
Twelve ounces for $0.35 is the better deal.

15. ($4.84 ÷ 25.5) × 32 = $6.0737. A quart of olive oil would cost $6.07.

16. a. 128 oz ÷ 17 = 7.53 oz Each person would get about 7 1/2 ounces of ice cream.
b. Each person would get 5 scoops of ice cream.

17. (10 × 16 oz) ÷ 7 = 22.85714 oz. = 1 lb 6.85714 oz. Each person would get 1 lb 7 oz.

Convert Metric Measuring Units, p. 127

1.

a. 3 cm = 3/100 m = 0.03 m	b. 2 cg = 2/100 g = 0.02 g
5 mm = 5/1000 m = 0.005 m	6 ml = 6/1000 L = 0.006 L
7 dl = 7/10 L = 0.7 L	1 dg = 1/10 g = 0.1 g

2.

a. 3 kl = 3,000 L	b. 2 dam = 20 m	c. 70 km = 70,000 m
8 dag = 80 g	9 hl = 900 L	5 hg = 500 g
6 hm = 600 m	7 kg = 7,000 g	8 dal = 80 L

3.

a. 3,000 g = 3 kg	b. 0.01 m = 1 cm	c. 0.04 L = 4 cl
800 L = 8 hl	0.2 L = 2 dl	0.8 m = 8 dm
60 m = 6 dam	0.005 g = 5 mg	0.007 L = 7 ml

4. a. 0.04 meters = 4 cm   b. 0.005 grams = 5 mg   c. 0.037 meters = 37 mm
   d. 400 liters = 4 hl   e. 0.6 meters = 6 dm   f. 2,000 meters = 2 km
   g. 0.206 liters = 206 ml   h. 20 meters = 2 dam   i. 0.9 grams = 9 dg

5. a. 45 cm = 0.45 m   b. 65 mg = 0.065 g   c. 2 dm = 0.2 m
   d. 81 km = 81,000 m   e. 6 ml = 0.006 L   f. 758 mg = 0.758 g
   g. 2 kl = 2,000 L   h. 8 dl = 0.8 L   i. 9 dag = 90 g

6. a. 12.3 m

		1	2.	3		
km	hm	dam	m	dm	cm	mm

b. 78 mm

			0.	0	7	8
km	hm	dam	m	dm	cm	mm

c. 56 cl

			0.	5	6	
kl	hl	dal	l	dl	cl	ml

d. 9.83 hg

	9	8	3.			
kg	hg	dag	g	dg	cg	mg

7.

	m	**dm**	**cm**	**mm**
a. 12.3 m	12.3	123	1230	12300
b. 78 mm	0.078	0.78	7.8	78
	L	**dl**	**cl**	**ml**
c. 56 cl	0.56	5.6	56	560
	g	**dg**	**cg**	**mg**
d. 9.83 hg	983	9830	98300	983000

Convert Metric Measuring Units, cont.

8. a. 560 cl = 5.6 L
   b. 0.493 kg = 49.3 dag
   c. 24.5 hm = 245,000 cm
   d. 491 cm = 4.91 m
   e. 35,200 mg = 35.2 g
   f. 32 dal = 32,000 cl
   g. 0.483 km = 4830 dm
   h. 0.0056 km = 560 cm
   i. 1.98 hl = 1980 dl
   j. 9.5 dl = 0.95 L

9. a. 13 cm b. 45 mm
   c. 0.92 m d. 2.4 m
   e. 1.70 m f. 1.34 m

10. a. The books weigh 1200 g + 1040 g + 520 g + 128 g = 2,888 g, or 2.888 kg.
    b. The total volume of the containers is 1.4 L + 2.25 L + 0.55 L + 0.24 L + 0.4 L = 4.84 L.

11. You can fill the 4-ml dropper 0.200 ÷ 0.004 = 50 times from the 2-dl bottle.

12. The patient will have received 2 grams of medicine in ten days. The 70-kg patient receives 3 mg/kg × 70 kg = 210 mg per day. Two grams is 2,000 mg. Nine days is not quite enough time, since 9 × 210 mg = 1,890 mg. So it is the tenth day that the patient finishes receiving the 2 g of medicine.

Convert Between Customary and Metric, p. 130

1. a. 2.5 cm b. 30 cm c. 1.5 km d. 1 L
   e. 4 lb f. 3 ft g. 240 ml h. 4 L

2.

a. 1 cm < 1 in	b. 1 L > 1 qt	c. 1 kg > 1 lb	d. 1 g < 1 oz
e. 4 in < 20 cm	f. 5 kg < 20 lb	g. 3 gal > 2 L	h. 7 m > 4 ft

3.

a. 1 cm = 0.39 in 25 cm = 9.84 in	b. 1 m = 1.09 yd 5.4 m = 17.72 ft	c. 2 L = 2.11 qt 4.6 L = 4.86 qt	d. 5 kg = 11.00 lb 0.568 kg = 1.25 lb
e. 5 in = 12.70 cm 10 in = 25.40 cm	f. 30 ft = 9.14 m 22 ft = 6.71 m	g. 1 gal = 3.78 L 3 1/2 qt = 3.31 L	h. 75 lb = 34.09 kg 8.5 lb = 3.86 kg

4. (c) 90 km/h. (Miles are *bigger* than kilometers, so there will be *more* kilometers. So 55 mi/hr × 1.6 km/mi = 88 km/hr.)

5. In ounces:

   The 24-oz bottle costs $6.75 ÷ 24 = $0.28125 per ounce.
   1 qt is 0.946 liters, so 1 liter = 1/0.946 = 1.05708 quarts. In ounces, this is 32 × 1.05708 = 33.826638 ounces.

   The 1-liter bottle costs $9.25 ÷ 33.826638 = $0.27345 per ounce.
   The 1-liter container is the better buy.

   In liters:

   24 ounces is 3/4 of a quart. 1 qt is 0.946 liters, so 3/4 qt = (3/4) × 0.946 L = 0.7095 liters.
   The 24-oz bottle costs $6.75 ÷ 0.7095 = $9.5137 per liter, and the 1-liter bottle costs $9.25 per liter.
   The latter is the better deal.

6. a. No, because 64 oz is equivalent to 2 qt, and a quart is smaller than a liter.
   64 oz = 8 C = 4 pt = 2 qt = 2 × 0.946 L = 1.892 L.
   b. The 64-oz container is 2.2 L − 1.892 L = 0.308 L less than the 2.2-liter container.

7. From the lightest to the heaviest: Angela (56 kg ≈ 123 lb), Theresa (128 lb), Elizabeth (60 kg = 132 lb), and Judy (137 lb).

8. If a marathon is 26.21875 miles (*i.e.*, 26 miles and 385 yards), then a half marathon is 26.21875 mi ÷ 2 × 1.6093 km/mi = 21.097 km, or 21 km and 97 m.

Mixed Review, p. 132

1. Ten to the power of eight is equal to a hundred million. $10^8 = 100{,}000{,}000$.

2. $3 \times 10^6 + 5 \times 10^5 + 4 \times 10^2 + 8 \times 10^1$

3.

a. 213 · 5,829 Estimation: 200 · 5,900 = 1,180,000 Exact: 1,241,577 Error of estimation: 61,577	b. 435,212 ÷ 993 Estimation: 435,000 ÷ 1,000 = 435 Exact: 438.28 Error of estimation: 3.28

4.

c	$c + \frac{2c}{5}$
10	$10 + \frac{2 \cdot 10}{5} = 10 + 4 = 14$
15	$15 + \frac{2 \cdot 15}{5} = 15 + 6 = 21$

c	$c + \frac{2c}{5}$
20	$20 + \frac{2 \cdot 20}{5} = 20 + 8 = 28$
25	$25 + \frac{2 \cdot 25}{5} = 25 + 10 = 35$

5. $5.1 \div 3 \times 10 = 17$. A full gas tank holds 17 gallons.

6. $98 ÷ 7 × 6 + $98 = $182. The total cost for both printers was $182.

7.

a. $\frac{15 + 150}{5} = 33$	b. $\frac{5}{15 + 5} = \frac{1}{4}$	c. $\frac{380 + 10}{12 - 9} = 130$

8. a. $(t - 1)^2$ b. $9 - x$
 c. $7 + S$ d. $8(4 + x + 2)$
 e. $x^2 \div (x + 1)$

9. a. $24 - 11 = 13$
 b. $(3/5) \cdot 7 = 21/5 = 4\ 1/5$

10.

a. x^5	b. $2p + 2$
c. $10 \cdot x^3$	d. $8z$
e. $3f + 2x$	f. $3s + 10$

11. a. $t \geq 18$
 b. $p \leq \$40$
 c. $a > 12$

12. 3, 4, 5, 6

13. a. $3(5x + 6) = 15x + 18$
 b. $2(8x + 2 + y) = 16x + 4 + 2y$

14.

a. $x + 78 = 412$ $x = 412 - 78$ $x = 334$	b. $\frac{x}{9} = 600$ $x = 9 \cdot 600$ $x = 5{,}400$	c. $y - 5 = 12 + 18$ $y = 30 + 5$ $y = 35$

Chapter 3 Review, p. 134

1. a. three ten-thousandths = 0.0003
   b. 39234 hundred-thousandths = 0.39234
   c. 4 millionths = 0.000004
   d. 2 and 5 thousandths = 2.005

2. a. 0.00039 = 39/100,000
   b. 0.0391 = 391/10,000
   c. 4.0032 = 40,032/10,000

3. a. 0.75 b. 1.4 c. 0.85 d. 0.44

4.

Organism	Size (fraction)	Size (micrometers)	Size (decimal)
amoeba proteus	$\frac{600}{1,000,000}$ meters	600 micrometers	0.0006 m
protozoa	from $\frac{10}{1,000,000}$ to $\frac{50}{1,000,000}$ m	from 10 to 50 micrometers	from 0.00001 to 0.00005 m
bacteria	from $\frac{1}{1,000,000}$ to $\frac{5}{1,000,000}$ m	from 1 to 5 micrometers	from 0.000001 to 0.000005 m

5. a. 0.000526 < 0.0062 < 0.0256 b. 0.000007 < 0.00008 < 0.000087

6.

	0.37182	0.04828384	0.39627	0.099568
To the nearest hundredth:	0.37	0.05	0.40	0.10
To the nearest ten-thousandth:	0.3718	0.0483	0.3963	0.0996

7. a. 0.024 b. 0.75 c. 3.043

8. a. 2.1 – 1.09342 = 1.00658
   b. 17 + 93.1 + 0.0483 = 110.1483

9. a. 0.1 + 0.04 = 0.14
   b. 0.01 + 0.04 = 0.05
   c. 0.0001 + 0.04 = 0.0401

10.

a. 0.48 ÷ 6 = 0.08 6 × 0.08 = 0.48	b. 1.5 ÷ 0.3 = 5 0.3 × 5 = 1.5	c. 0.056 ÷ 0.008 = 7 0.008 × 7 = 0.056

11.

a. 3 × 0.006 = 0.018	b. 0.2 × 0.6 = 0.12	c. 0.9 × 0.0007 = 0.00063

12. 327 × 4 is 1,308. In the calculation 32.7 × 0.004, the decimals have one and three decimal digits, or four decimal digits in total. So we take 1,308 and make it have four decimal digits, so it becomes 0.1308.

13. a. Estimate: 9 × 0.06 = 0.54. b. Exact: 8.9 × 0.061 = <u>0.5429</u>

14. a. 0.03 b. 0.12 c. 0.05

15. a. $p = 225$
    b. $x = 173.33$
    c. $y = 0.324$

16. There is 4 m − (7 × 0.56 m) = 4 m − 3.92 m = <u>0.08 m</u> left.

17.

a. $10^6 \times 21.7 = 21,700,000$	b. 100 × 0.00456 = 0.456
c. 2.3912 ÷ 1,000 = 0.0023912	d. $324 \div 10^5 = 0.00324$
e. $10^5 \times 0.003938 = 393.8$	f. $0.7 \div 10^4 = 0.00007$

18. $\frac{a}{b} + 1 = 3.585$

19. a. 14.1 b. 0.007 c. 0.444 d. 0.455

20.

Prefix	Meaning	Units - length	Units - mass	Units - volume
centi-	hundredth = 0.01	centimeter (cm)	centigram (cg)	centiliter (cl)
deci-	tenth = 0.1	decimeter (dm)	decigram (dg)	deciliter (dl)
deca-	ten = 10	decameter (dam)	decagram (dag)	decaliter (dal)
hecto-	hundred = 100	hectometer (hm)	hectogram (hg)	hectoliter (hl)

21. a. 34 dl = 3.4 L
b. 89 cg = 0.89 g
c. 16 kl = 16,000 L

22. a. 2.7 L = 27 dl = 270 cl = 2700 ml
b. 5,600 m = 5.6 km = 56,000 dm = 560,000 cm
c. 676 g = 6,760 dg = 67,600 cg = 676,000 mg

23. The total capacity is 6 × 0.35 L + 2 × 2 L + 3 × 0.9 L = 2.1 L + 4 L + 2.7 L = 8.8 L.

24. a. 56 oz = 3.5 lb c. 2.7 gal = 10.8 qt e. 0.48 mi = 2,534.4 ft
b. 134 in = 11.17 ft d. 0.391 lb = 6.26 oz f. 2.45 ft = 2 ft 5.4 in

25. The ribbons will measure 500 ft ÷ 230 = 2.17 ft = 2 ft 2 in.

26. The 40 yards of rope cost $15.99 / 40 yd = $0.40 per yard, and the 100 meters cost $40 ÷ 100 = $0.40 per meter. A meter is longer than a yard so the 100 meters is the better deal.

27. Since 3.2 + 3.1 + 3.4 + 3.1 + 3.5 + 2.9 + 2.7 + 2.7 + 3.0 + 3.0 + 3.1 + 3.4 + 3.2 + 2.8 + 2.8 + 2.9 + 3.6 + 3.4 + 2.9 + 3.4 + 3.1 = 65.2, the average length of the 21 tadpoles was 65.2 / 21 = 3.1 cm.
(We give the answer to tenths because that was the accuracy of all of the data.)

Chapter 4: Ratios

Ratios and Rates, p. 141

1.

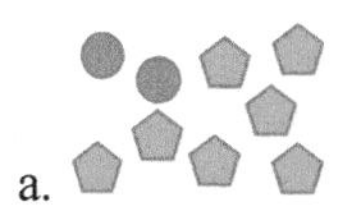 a. The ratio of circles to pentagons is 2:7. The ratio of pentagons to all shapes is 7:9. $\frac{7}{9}$ of the shapes are pentagons.	 b. The ratio of hearts to stars is 5:10. The ratio of stars to all shapes is 10:15. $\frac{10}{15}$ of the shapes are stars.

2. a. Answers will vary. For example: 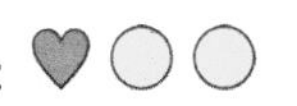 or 
   b. 1:2

3. a. There would be six circles. b. There would be ten circles. c. There would be 200 circles.

4. a. The ratio of diamonds to triangles is 9:6 or 3:2. There are 3 diamonds to every 2 triangles.
   b. The ratio of pentagons to circles is 8:12 or 2:3. There are 3 circles to every 2 pentagons.

5. a. 
   b. There are 6 hearts to 15 squares, 6:15 = 2:5.
   c. The ratio of all shapes to hearts is 21:6 = 7:2.

6. a. 15:20 = 3:4 b. 16:4 = 4:1 c. 25:10 = 5:2 d. 13:30 (is already in lowest terms)

7.

a. $\frac{5}{2} = \frac{20}{8}$	b. 3:4 = 9:12	c. 16:18 = 8:9	d. $\frac{5}{1} = \frac{20}{4}$
e. 2 to 100 = 1 to 50	f. 24 to 40 = 3 to 5	g. 5:100 = 1 to 20	

8.

a. 2 kg and 400 g $\frac{2 \text{ kg}}{400 \text{ g}} = \frac{2{,}000 \text{ g}}{400 \text{ g}} = \frac{2{,}000}{400} = \frac{5}{1}$	b. 200 ml and 2 L $\frac{200 \text{ ml}}{2 \text{ L}} = \frac{200 \text{ ml}}{2{,}000 \text{ ml}} = \frac{200}{2{,}000} = \frac{1}{10}$
c. 4 cups and 2 quarts $\frac{4 \text{ c}}{2 \text{ qt}} = \frac{4 \text{ c}}{8 \text{ c}} = \frac{4}{8} = \frac{1}{2}$	d. 800 m and 1.4 km $\frac{800 \text{ m}}{1.4 \text{ km}} = \frac{800 \text{ m}}{1{,}400 \text{ m}} = \frac{800}{1{,}400} = \frac{4}{7}$
e. 120 cm and 1.8 m $\frac{120 \text{ cm}}{1.8 \text{ m}} = \frac{120 \text{ cm}}{180 \text{ cm}} = \frac{120}{180} = \frac{2}{3}$	f. 3 ft 4 in and 1 ft 4 in $\frac{3 \text{ ft } 4 \text{ in}}{1 \text{ ft } 4 \text{ in}} = \frac{40 \text{ in}}{16 \text{ in}} = \frac{40}{16} = \frac{5}{2}$

9. a. Annie walks at the rate of 3 km per 30 min, which is 1 km per 10 min or 6 km per hour.
   b. The rate of teachers to students is 5:60 or 1:12.
   c. Rice cost \$4.50/3 kg or \$1.50/kg.
   d. The car travels 66 miles per 3 gallons or 22 miles per 1 gallon of gasoline.

Ratios and Rates, cont.

10.

a. $\frac{2\text{ cm}}{30\text{ min}} = \frac{1\text{ cm}}{15\text{ min}} = \frac{3\text{ cm}}{45\text{ min}}$	b. $\frac{\$72}{8\text{ hr}} = \frac{\$9}{1\text{ hr}} = \frac{\$90}{10\text{ hr}}$
c. $\frac{1/4\text{ mi}}{10\text{ min}} = \frac{1\ 1/2\text{ mi}}{1\text{ hr}} = \frac{7\ 1/2\text{ mi}}{5\text{ hr}}$	d. $\frac{\$84.40}{8\text{ hr}} = \frac{\$21.10}{2\text{ hr}} = \frac{\$105.50}{10\text{ hr}}$

11. a. $44:4 hr = $11:1 hr
    b. $30:8 kg = $15:4 kg
    c. 420 km:8 hr = 105 km:2 hr
    d. 16 apples for $12 = 4 apples:$3

12.

Pencils	Dollars
1	0.24
2	0.48
3	0.72
6	1.44
7	1.68
8	1.92

13.

Miles	21	42	63	84	105	210	315	1,050
Gallons	1	2	3	4	5	10	15	50

14.

Km	10	20	80	100	150	200	500
Minutes	7 1/2	15	60	75	112 1/2	150	375

15. Five pairs of scissors would cost $17.50.

Scissors	1	2	3	4	5	6
Cost	$3.50	$7.00	$10.50	$14	$17.50	$21

16. Mark can type 540 words in 12 minutes.

Unit Rates, p. 145

1. Answers will vary. Please check the student's work. Examples: John drove 50 miles per hour (50 mph). Corn costs $0.50/lb.

2. a. $3 for 1 cup b. 30 miles in 1 hour or 30 mph (the common abbreviation for miles per hour)
   c. 2,000 people : 1 doctor

3. It is customary to omit the 1 from the unit ratio. For example, $5 per 1 lunch is written as $5 per lunch.
   a. 350 sq. ft. / gallon b. 4 megabits per second c. 2 1/2 students / calculator or 1 calculator per 2.5 students
   d. 1 3/4 tsp per cup e. $5 per lunch f. $22.67 / mattress

4. His unit rate is 33 ft per 3 hours = 11 ft per hour. To paint the 98 feet remaining will take 98 ft ÷ 11 ft/hr = 8.909 hours. To convert this time into hours and minutes, we convert the 0.9 hours into minutes: 0.9 hours · 60 minutes/hour = 54 minutes. So, painting at the same speed, it will take Jack 8 hours 54 minutes more. Note: In real life, rounding to nine hours is probably accurate enough. He will take the last six minutes collecting his brushes and paint cans.

5. a. Jerry could mow 20 lawns of the same size in 35 hours. (Five times as many lawns as he can in seven hours.)
   b. 7 hours per 4 lawns = 7/4 hours per lawn = 1.75 hours or 1 h 45 min per lawn.

6. The unit rate is 4,400 km : 5 h = 880 km : 1 h. In eight hours, the airplane can travel 8 · 880 km = 7,040 km.

7. A recipe has a ratio of 3/8 cups of sugar per 1 cup of flour. 1 1/2 C to 4 C = 3/4 C to 2 C = 3/8 C to 1 C.

8. The instructions are equivalent to 2.5 lb fertilizer per 1,000 sq. ft.
   She would need to apply 17.5 pounds of fertilizer to a lawn that is 7,000 sq. ft.

9. 45 L per 700 m^2 is 1 liter per 15.5555...m^2. You can also write this rate the other way, as it may be easier to see what to do to calculate the unit rate: 700 m^2 per 45 L = 15.5555 m^2 per 1 L. Six liters of paint would cover six times that, or 93.33 m^2. Or, you can use a table to help you reason it out:

Paint	45 L	9 L	3 L	6 L
Wall area	700 m^2	140 m^2	46.67 m^2	93.33 m^2

Using Equivalent Rates, p. 147

1.

a. $\frac{15\text{ km}}{3\text{ hr}} = \frac{5\text{ km}}{1\text{ hr}} = \frac{1.25}{15\text{ min}} = \frac{3.75}{45\text{ min}}$	b. $\frac{\$6}{45\text{ min}} = \frac{\$2}{15\text{ min}} = \frac{\$8}{1\text{ hr}} = \frac{\$14}{1\text{ hr }45\text{ min}}$
c. $\frac{3\text{ in}}{8\text{ ft}} = \frac{0.75\text{ in}}{2\text{ ft}} = \frac{4.5\text{ in}}{12\text{ ft}} = \frac{7.5\text{ in}}{20\text{ ft}}$	d. $\frac{115\text{ words}}{2\text{ min}} = \frac{57.5\text{ words}}{1\text{ min}} = \frac{172.5\text{ words}}{3\text{ min}}$

2. a. $\frac{8\text{ miles}}{14\text{ minutes}} = \frac{4\text{ miles}}{7\text{ minutes}} = \frac{36\text{ miles}}{63\text{ minutes}}$

   b. Jake can ride 20 miles in 35 minutes.

3. a. You would need 2.4 gallons of gasoline.

   $\frac{50\text{ miles}}{2\text{ gallons}} = \frac{5\text{ miles}}{0.2\text{ gallon}} = \frac{60\text{ miles}}{2.4\text{ gallons}}$

   b. At 25:1 a car can travel 375 miles on 15 gallons of gasoline.

4. It would cost $1.98 for 22 erasers.

Cost (C)	$0.09	$0.18	$0.90	$1.80	$1.98
Erasers (E)	1	2	10	20	22

5. You could buy 15 erasers for $1.35.

6. He makes baskets at a rate of 3 baskets to every 4 shots so he can expect to make 150 baskets with 200 shots. If you use the table, there are many ways to fill it in, and it doesn't need to be completely filled in, in order to solve the problem. All the table is for is to help the student write down several equivalent rates. For example:

baskets	9	3	15	150			
shots	12	4	20	200			

7. a.

C	3	6	9	12	15	18	21	24	27	30
p	1	2	3	4	5	6	7	8	9	10

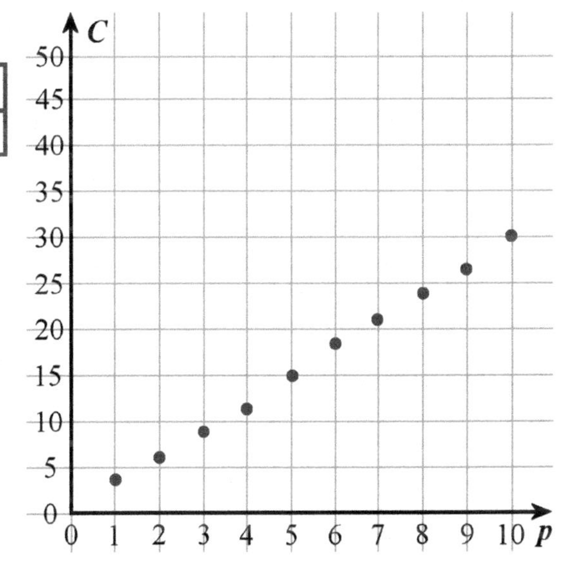

   c. $C = 3p$

8. a. It would cost $7.80 for 52 pencils.

Cost	$0.15	$1.50	$4.50	$6.00	$7.50	$7.80
Pencils	1	10	30	40	50	52

   b. $C = 0.15p$ or $C = (3/20)p$

9. a.

t	0.8	1.6	2.4	3.2	4	4.8	5.6	6.4	7.2	8
j	1	2	3	4	5	6	7	8	9	10

c. Kate needs 25 jars for 20 liters of tea.

d. There will be 12.8 liters of tea in 16 jars.

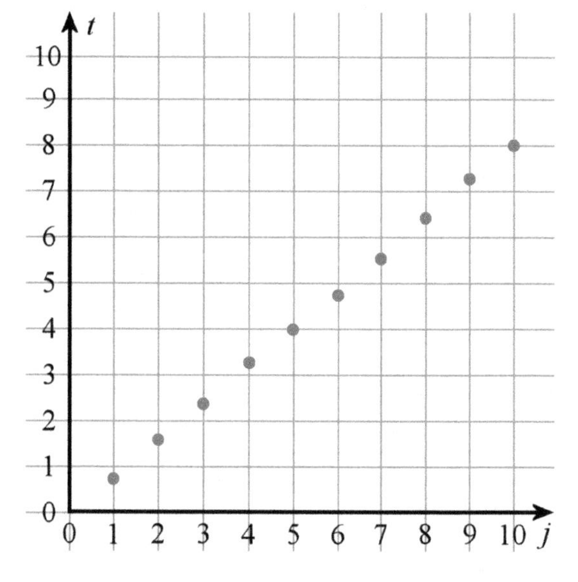

10. a.

d	80	160	240	320	400	480	560	640	720	800
h	1	2	3	4	5	6	7	8	9	10

b. $d = 80h$ or $h = d/80$
c. See the grid on the lower right.

11.

d	60	120	180	240	300
h	1	2	3	4	5

d	360	420	480	540	600
h	6	7	8	9	10

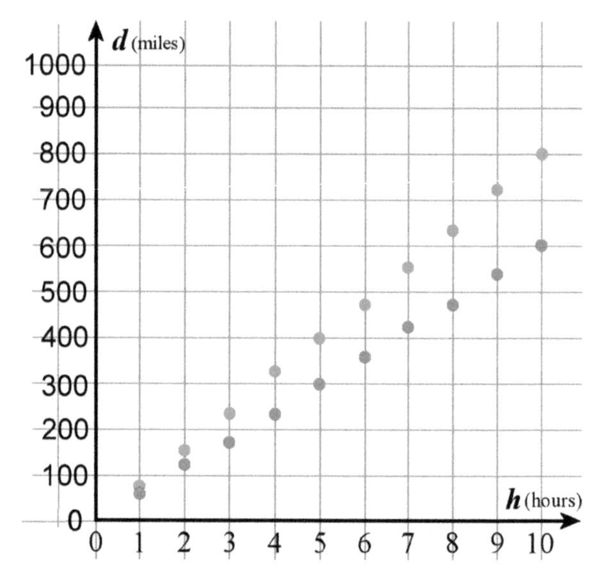

12. You can tell by noting which line of dots rises more steeply (faster) on the grid.

Using Equivalent Rates, cont.

13. Person 1 (red dot)

d	2/3	1 1/3	2	2 2/3	3 1/3	4	4 2/3	5 1/3	6	6 2/3
t	10	20	30	40	50	60	70	80	90	100

Person 2 (blue dot)

d	1/2	1	1 1/2	2	2 1/2	3	3 1/2	4	4 1/2	5
t	10	20	30	40	50	60	70	80	90	100

a. Person 1 walks at the speed of 4 miles per hour.

b. Person 2 walks at the speed of 3 miles per hour.

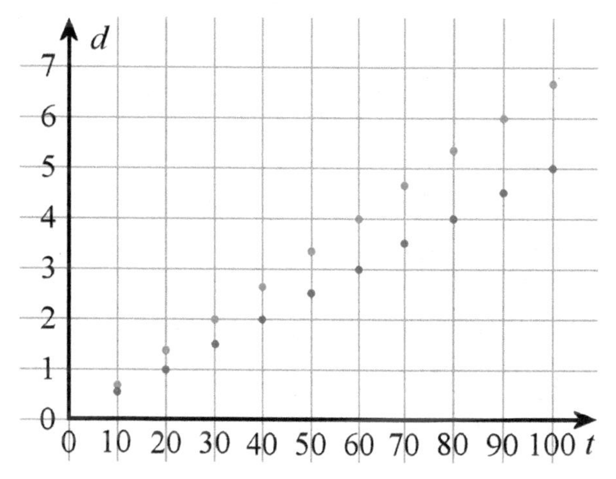

14. 240 mi per 3 h = 80 mi per 1 h and 490 mi per 7 h = 70 mi per 1 h. Train one is faster.

15. \$45 : 8 bottles = \$5.63 : 1 bottle and \$34 : 6 bottles = \$5.67 : 1 bottle.
Eight bottles of shampoo for \$45 is the better deal, and is about 4 cents cheaper per bottle.

16. a. 640 people *out of* 1000 people = 16 people *out of* 25 people
b. 64 people out of 100 people
c. 144 people out of 225 people

Ratio Problems and Bar Models 1, p. 151

1.
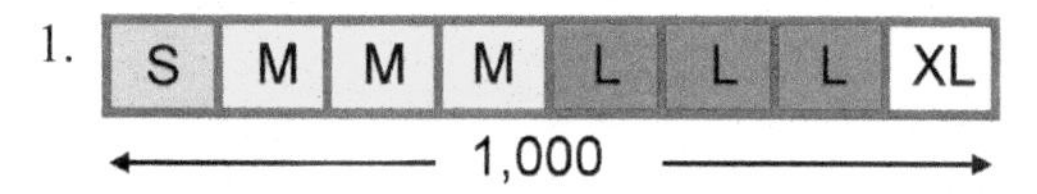

a. Small (S) shirts are in a ratio of 1 : 8 to the total number of shirts.
b. For each 1,000 shirts made there are 125 S, 375 M, 375 L, and 125 XL size shirts.

2. a. To make 3 liters of juice, each of the 7 parts would be 3,000 ml ÷ 7 = 430 ml or 0.43 L (rounded to the nearest 10 ml), so you want 2 × 0.43 L = 0.86 L of concentrate and 5 × 0.43 L = 2.14 L of water.

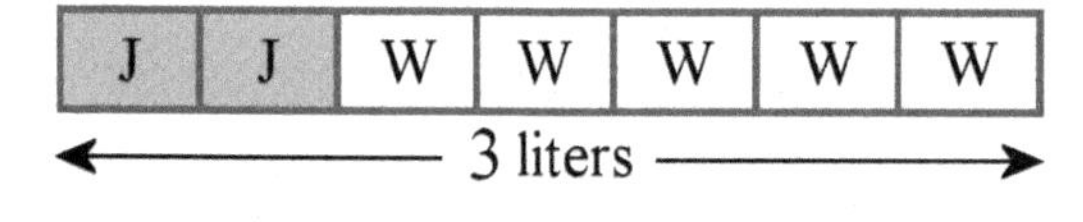

b. This 1/2 liter of concentrate corresponds to two parts, so each part is 1/4 L. Therefore, add 5 parts of water, which is 5 × (1/4)L = 1.25 L of water. This makes 1.75 L of diluted juice.

3. There are 8 parts, and each part is worth \$250.00 / 8 = \$31.25. Greg earned three parts, or 3 × \$31.25 = \$93.75, and Matthew earned five parts, or 5 × \$31.25 = \$156.25.

4. If Ann received \$80 for her two parts, then one part is \$40.
a. For her five parts Shelly received 5 × \$40 = \$200.
b. The total salary was all seven parts, or 7 × \$40 = \$280.

Ratio Problems and Bar Models 1, cont.

5. The ratio *white* : *translucent* : *colored* is 1 : 3 : 5, so there are 1 + 3 + 5 = 9 parts in total. If there are 75 colored marbles, since colored marbles make up 5 parts of the total, then one part consists of 75 ÷ 5 = 15 marbles.
   a. The bag has all the marbles, a total of nine parts, or 9 × 15 = 135 marbles.
   b. The white marbles make up only one part of the total, so there are 15 of them.
   c. The translucent marbles make up three parts of the total, so there are 3 × 15 = 45 translucent marbles.

6. The ratio of roses is *white* : *red* : *yellow* : *pink* = 2 : 1 : 2 : 3, so there are 2 + 1 + 2 + 3 = 8 parts. The 69 pink roses make up 3 parts of the total, so each part is 69 ÷ 3 = 23 roses. The total arrangement contains all 8 parts, or 8 × 23 = 184 roses.

7. a. The school currently has a ratio of teachers to students of 24 to 600. To reduce this ratio to lowest terms, divide each of the numbers by the largest factor that they share (24). The ratio in lowest terms is thus 1 : 25.
   b. To maintain a ratio of teachers to students of 1 : 25, a school with 800 students would need 800 ÷ 25 = 32 teachers. So the school would need to hire 32 − 24 = 8 more teachers.

8. a. The ratio of Nissans to Fords is 4:5.
   b. The ratio of Fords to all cars is 5:9.
   c. There is a total of 4 + 5 = 9 parts, so each part consists of 450 ÷ 9 = 50 cars. So the 5 parts that are Fords consist of 5 × 50 = 250 Fords.

9. a.

J	J	J	R	R	R	R	R

   b. The ratio of Joe's money to Rita's money was 3:5.
   c. Rita gave Joe 1 of her 5 parts of money, so she now has 1 part less, and Joe has 1 part more. So the ratio of Joe's money to Rita's money is now 4:4 = 1:1.

10. The ratio of Britney's weight to Joan's weight is 5 to 6 (why?). So their total weight of 121 kg consists of 11 parts, each of which is 121 ÷ 11 = 11 kg. Britney weighs 5 parts of the total, or 5 × 11 = <u>55 kg</u> (about 121 lbs.).

11. a.

F	F	G	G	G	G	G	G	G	G	G

One third of Greg's 9 parts is 3 parts.

   b.

F	F	F	F	F	G	G	G	G	G	G

   c. Greg's 6 parts consist of 120 marbles, so each part contains 120 ÷ 6 = 20 marbles, and Fred has 5 × 20 = 100 marbles now.

Ratio Problems and Bar Models 2, p. 154

1. a. The ratio of red phones to silver phones is 2 : 7, so there are 2 + 7 = 9 parts. The five more parts of silver phones consist of 300 phones, so each part consists of 300 ÷ 5 = 60 phones.
   b. The crate contains all nine parts, so it has 9 × 60 = 540 phones.
   c. Since two parts of the phones are red, 2 × 60 = 120 are red.

R	R					
		300				
S	S	S	S	S	S	S

2. Eric has 3 parts of the cards, and Erica has 4 parts. The 1 part more that Erica has consists of 14 cards, so each of the other parts will also contain 14 cards. So Eric has 3 × 14 = 42 phone cards.

3. The ratio of the length of Mr. Short's life to Mr. Long's was 3 : 7, so there are 3 + 7 = 10 parts in all. The 44 years longer that Mr. Long lived make up 4 of those parts, so each part lasted 44 ÷ 4 = 11 years. Thus Mr. Long lived a total of 7 × 11 = 77 years.

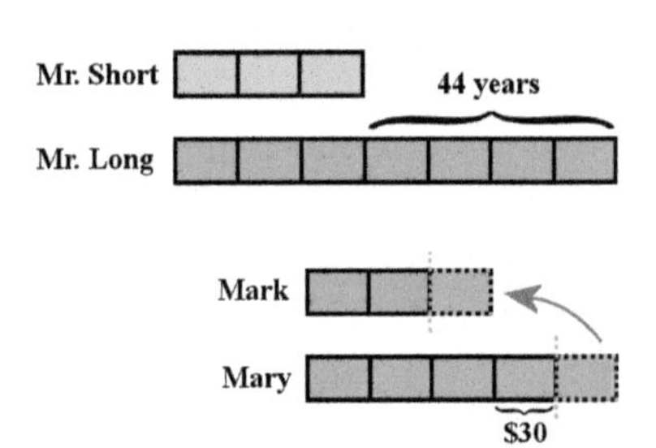

4. Mark and Mary's shares were in a ratio of 2:5, so there were 2 + 5 = 7 parts in all. One fifth of Mary's 5 parts consists of 1 part, so after Mary gave Mark that part, there were still 7 parts in all, and their shares were in a ratio of 3:4. Mary ended up with one only part more, and it was worth $30. So the total sum of money was 7 × $30 = $210.

5. The total length minus the difference in length is 180 cm − 50 cm = 130 cm. The shorter board is half of that result, or 130 cm ÷ 2 = 65 cm. The longer board is the length of the shorter board plus the difference in their lengths, or 65 cm + 50 cm = 115 cm.

6. Taking away the eight-crayon difference leaves 56 − 8 = 48 crayons. Margie's sister got half of that number, or 24 crayons. So Margie got 24 + 8 = 32 crayons.

7. Taking away the $1 difference from the total bill leaves $71.40. Dividing that amount into the 6 DVDs gives $71.40 ÷ 6 = $11.90. So the more expensive DVD cost $11.90 + $1.00 = $12.90.

8. The ratio of the length of the shortest necklace to that of the longest is 5:7, so the difference of 18 cm is 7 − 5 = 2 parts, and each part is thus 18 ÷ 2 = 9 cm. So the length of the shortest necklace is 5 parts, or 5 × 9 cm = 45 cm, the length of the mid-length necklace is 6 parts, or 6 × 9 cm = 54 cm, and the length of the longest necklace is 7 parts, or 7 × 9 cm = 63 cm.

9. After Alice and Cindy both use 2 parts of their money, the ratio of Alice's money to Cindy's money is 1:3. Therefore, 2 parts of Alice's money must equal $30, and one part equals $15. So at first Alice had 3 × $15 = $45.

10. The 1/4 of cookies in jar 1 that Mark moves corresponds to exactly one block. Afterwards, jar 2 has 6 "blocks" of cookies, and jar 1 has 3 "blocks". The difference of 24 cookies corresponds to 3 blocks. So, one block is 8 cookies. There is a total of 9 × 8 = 72 cookies.

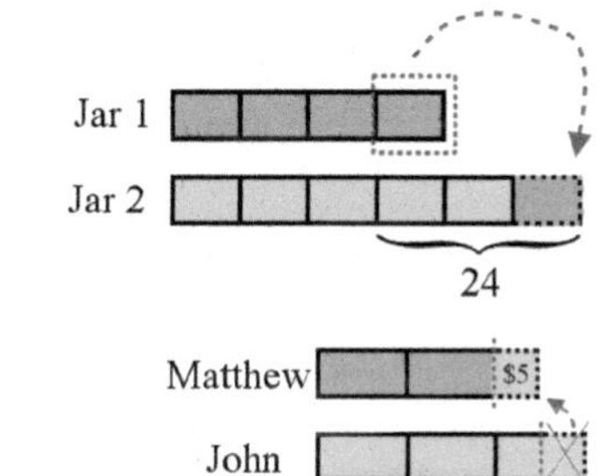

11. In this problem, you can first draw the diagram with the ratio 2:3. Since the two boys will have the same amount of money afterwards, this means that the $5 that John gave to Matthew is 1/2 of a block in the diagram. Thus, one block is $10, and John had 3 × $10 = $30 in the beginning.

12. This version may be easier to solve by starting at the end and working toward the beginning. The final ratio of 7:8 gives a total of 15 parts to our diagram. Then if Matthew gave the $5 back to John, we would get back to the initial ratio of 2:3, but we still need to have those 15 parts. A ratio of 2:3 is only 5 parts. But we can change 5 parts into 15 parts if we multiply by three to make the equivalent ratio 6:9. Now we have a model where moving one part turns the ratio 6:9 into 7:8, and that one part that moves is the $5. So in the beginning John's nine parts were 9 × $5 = $45.

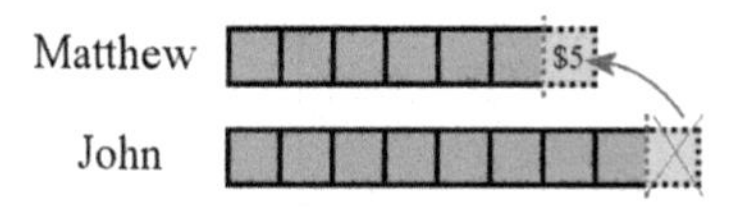

13. At first, Annie has 5 parts of the total and Michelle has 8. Michelle sells half of hers, so she has 4 parts left. Annie sells 21 roses and ends up with 1/2 as many roses as Michelle, so Annie must have only 2 parts left. Therefore, Annie sold 3 of the original parts, and that was 21 roses. So, one part is 7 roses. Michelle had 8 × 7 roses = 56 roses in the beginning.

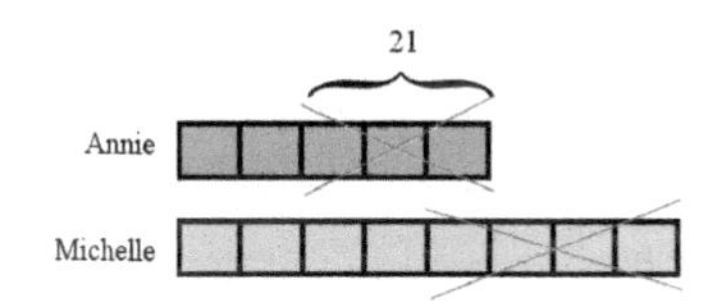

Michelle has four parts left, so Annie must have two parts left.

Aspect Ratio, p. 157

1. a. 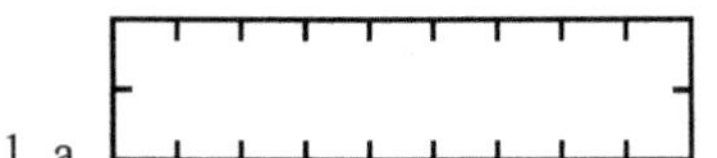
   b. The rectangle's aspect ratio is 9:2, so its perimeter consists of 9 + 2 + 9 + 2 = 22 parts, and each part is 220 cm ÷ 22 = 10 cm long. So the rectangle's width is 9 × 10 cm = 90 cm, and its height is 2 × 10 cm = 20 cm.

2. The rectangle's aspect ratio is 3:1, so its perimeter is made up of 3 + 1 + 3 + 1 = 8 parts. Each part is 120 mm ÷ 8 = 15 mm long. So its width is 3 × 15 mm = 45 mm, and its length is 1 × 15 mm = 15 mm.

3. a. The rectangle's width is 5 parts and its height is 2 parts, therefore the aspect ratio is *width*:*height* = 5:2.
   b. The rectangle's aspect ratio is *width*:*height* = 1:5.
   c. A square's aspect ratio is just *width*:*height* = 1:1.

4. a. Think of the height being 9 parts and the width being 4 parts. Then the width is 4/9 of the height, and the ratio of the door's width to its height is 4:9.
   b. Since the ratio *width*:*height* = 4:9, the width of 54 cm is 4 parts of the ratio. The height is 9 parts, or 54 cm ÷ 4 × 9 = 121.5 cm.

5. a. The aspect ratio is *width*:*height* = 6:9 = 2:3.
   b. The width is two parts, so each part is 20 in ÷ 2 = 10 in. The height is 3 parts, or 3 × 10 in = 30 in.

6. a. The simplified aspect ratios of all the windows are:

Size	*Simplified*
70 cm x 90 cm	7:9
80 cm x 100 cm	8:10 = 4:5
90 cm x 110 cm	9:11
100 cm x 120 cm	10:12 = 5:6

   b. *None* of the windows share the same aspect ratio. This means none of them have exactly the same shape.

7. a. The aspect ratio is *width*:*height* (length) = 2:1.
   b. In terms of the aspect ratio of 2:1, the perimeter is 2 + 1 + 2 + 1 = 6. If the perimeter measures 15 ft, then each part of the ratio is 15 ft ÷ 6 = 2.5 ft. So its width is 2 × 2.5 ft = 5 ft, and its length is 1 × 2.5 ft = 2.5 ft.
   c. From the answer to part (b), its area is *width* × *length* = 5 ft × 2.5 ft = 12.5 sq ft.

8. The aspect ratio of the first television is 16:9, so the perimeter consists of 16 + 9 + 16 + 9 = 50 parts, and each part has a length of 150 cm ÷ 50 = 3 cm.

   The aspect ratio of the second television is 4:3, so the perimeter consists of 4 + 3 + 4 + 3 = 14 parts, and each part is 150 cm ÷ 14 ≈ 10.714 cm ≈ 10.7 cm.

   a. The 16:9 television is 16 × 3 cm = 48 cm wide and 9 × 3 cm = 27 cm high.

   The 4:3 one is 4 × 10.714 cm = 42.856 cm ≈ 42.9 cm wide and 3 × 10.714 cm = 32.142 cm ≈ 32.1 cm high.

   b. The area of the 16:9 television is *width* × *height* = 48 cm × 27 cm = 1,296 cm^2.

   The area of the 4:3 model is 42.856 cm × 32.142 cm ≈ 1,377.48 cm^2 ≈ 1,377 cm^2.

   Therefore, the 4:3 television has the larger area.

9. If the area of a square is 49 sq. in. = 7 in × 7 in, then each of its sides must measure 7 in.
   a. The rectangle has dimensions 14 in by 7 in, so its aspect ratio is thus 14:7 = 2:1.
   b. The perimeter of the rectangle is 14 + 7 + 14 + 7 = 42 in.

Using Ratios to Convert Measuring Units, p. 159

1.

1 ft = 0.3048 m	1 ounce = 28.35 g	1 mi = 1,760 yd	1 m = 1.0936 yd
↓	↓	↓	↓
$\frac{1\text{ ft}}{0.3048\text{ m}} = 1$	$\frac{1\text{ oz}}{28.35\text{ g}} = 1$	$\frac{1\text{ mi}}{1{,}760\text{ yd}} = 1$	$\frac{1\text{ m}}{1.0936\text{ yd}} = 1$

2.

a. $79\text{ in} = 79\text{ in} \cdot 1 = 79\text{ in} \cdot \frac{2.54\text{ cm}}{1\text{ in}} = 79 \cdot 2.54\text{ cm} = 200.66\text{ cm} \approx 200.7\text{ cm}$

b. $56\text{ km} = 56\text{ km} \cdot 1 = 56\text{ km} \cdot \frac{1\text{ mi}}{1.6093\text{ km}} = \frac{56\text{ mi}}{1.6093} = 34.7977\text{ mi} \approx 34.8\text{ mi}$

c. $2.8\text{ mi} = 2.8\text{ mi} \cdot 1 = 2.8\text{ mi} \cdot \frac{1.6093\text{ km}}{1\text{ mi}} = 2.8 \cdot 1.6093\text{ km} = 4.50604\text{ km} \approx 4.5\text{ km}$

d. $4\text{ qt} = 4\text{ qt} \cdot 1 = 4\text{ qt} \cdot \frac{0.946\text{ L}}{1\text{ qt}} = 4 \cdot 0.946\text{ L} = 3.784\text{ L} \approx 3.8\text{ L}$

3.

a. $89\text{ cm} = 89\text{ cm} \cdot 1 = 89\text{ cm} \cdot \frac{1\text{ in}}{2.54\text{ cm}} = \frac{89\text{ in}}{2.54} = 35.0394\text{ in} \approx 35.0\text{ in}$

b. $15\text{ kg} = 15\text{ kg} \cdot 1 = 15\text{ kg} \cdot \frac{2.2\text{ lb}}{1\text{ kg}} = 15 \cdot 2.2\text{ lb} = 33\text{ lb}$

c. $78\text{ mi} = 78\text{ mi} \cdot 1 = 78\text{ mi} \cdot \frac{1.6093\text{ km}}{1\text{ mi}} = \frac{78 \cdot 1.6093\text{ km}}{1} = 125.529\text{ km} \approx 125.5\text{ km}$

d. $89\text{ ft} = 89\text{ ft} \cdot 1 = 89\text{ ft} \cdot \frac{0.3048\text{ m}}{1\text{ ft}} = 89 \cdot 0.3048\text{ m} = 27.1272\text{ m} \approx 27.1\text{ m}$

e. $365\text{ g} = 365\text{ g} \cdot 1 = 365\text{ g} \cdot \frac{1\text{ oz}}{28.35\text{ g}} = \frac{365\text{ oz}}{28.35} = 12.8748\text{ oz} \approx 12.9\text{ oz}$

4.

a. $5\text{ ft} = 5\text{ ft} \cdot \frac{2.54\text{ cm}}{1\text{ in}} \cdot \frac{12\text{ in}}{1\text{ ft}} = 5 \cdot 12 \cdot 2.54\text{ cm} = \approx 152\text{ cm}$

b. $24\text{ oz} = 24\text{ oz} \cdot \frac{1\text{ qt}}{32\text{ oz}} \cdot \frac{0.946\text{ L}}{1\text{ qt}} = \frac{24 \cdot 0.946\text{ L}}{32} = 0.7095 \approx 0.71\text{ L}$

c. $700\text{ yd.} = 700\text{ yd} \cdot \frac{0.9144\text{ m}}{1\text{ yd}} = (700\text{ yd} \cdot 0.9144\text{ m}/1\text{yd}) = 700 \cdot 0.9144\text{ m} = 640.08\text{ m} \approx 640.1\text{ m}$

d. $8\text{ kg} = 8\text{ kg} \cdot \frac{2.2\text{ lb}}{1\text{ kg}} \cdot \frac{16\text{ oz}}{1\text{ lb}} = 8 \cdot 2.2 \cdot 16\text{ oz} = 281.6 \approx 282\text{ oz}$

e. $371\text{ oz} = 371\text{ oz} \cdot \frac{28.35\text{ g}}{1\text{ oz}} = 371 \cdot 28.35\text{ g} = 10517.85 \approx 10{,}500\text{ g}$

f. $15\text{ pt} = 15\text{ pt} \cdot \frac{0.946\text{ L}}{1\text{ qt}} \cdot \frac{1\text{ qt}}{2\text{ pt}} = \frac{15 \cdot 0.946\text{ L}}{2} = 7.095 \approx 7.10\text{ L}$

Mixed Review, p. 163

1. 4,958/13 = 381 R5

2. 43 ÷ 9 = 4.778

3. a. 51,999,601 ≈ 52,000,000
   b. 109,999,339 ≈ 110,000,000

4.

a. $3 \times 0.25 = 0.75$ $4 \times 0.025 = 0.1$	b. $8 \times 0.08 = 0.64$ $100 \times 0.0008 = 0.08$	c. $1 \div 0.05 = 20$ $4 \div 0.05 = 80$	d. $0.99 \div 11 = 0.09$ $0.06 \div 0.001 = 60$

5. a. $10^5 \times 3.07 = 307000$	b. $10^4 \times 0.00078 = 7.8$
c. $12.7 \div 10^3 = 0.0127$	d. $5{,}600 \div 10^5 = 0.056$

6. The length of each side is $2y$.

7.

Expression	the terms in it	coefficient(s)	Constants
$2a + 3b$	$2a$ and $3b$	2 and 3	
$10s$	$10s$	10	
$11x + 5$	$11x$ *and* 5	11	5
$8x^2 + 9x + 10$	$8x^2$ and $9x$ and 10	8 and 9	10
$\frac{1}{6}p$	$\frac{1}{6}p$	$\frac{1}{6}$	1/6

8. a.

X	1	2	3	4	5	6	7	8	9
Y	9	8	7	6	5	4	3	2	1

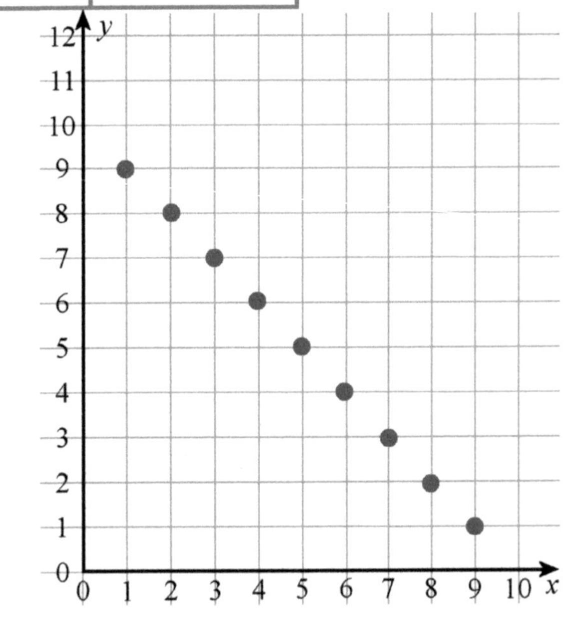

b. $X + Y = 10$ or $Y = 10 - X$.
c. In this problem, you could choose either one of the variables to be plotted on the horizontal axis. The most likely choice is X, though.

9. a. $5n$ b. $67 - y$ c. $\frac{8}{10}p$

10. a. 675.5 ÷ 0.3 = 2,251.667

b. $\frac{2}{7} = 0.286$

Chapter 4 Review, p. 165

1. a. $\frac{4}{3} = \frac{20}{15}$	b. 6:7 = 18:21	c. 4 to 30 = 2 to 15	d. $\frac{7}{3} = \frac{28}{12}$

2. a. $\frac{15}{35} = \frac{3}{7}$	b. $\frac{6}{16} = \frac{3}{8}$	c. 33:30 = 11:10	d. 9:12 = 3:4

3. a.

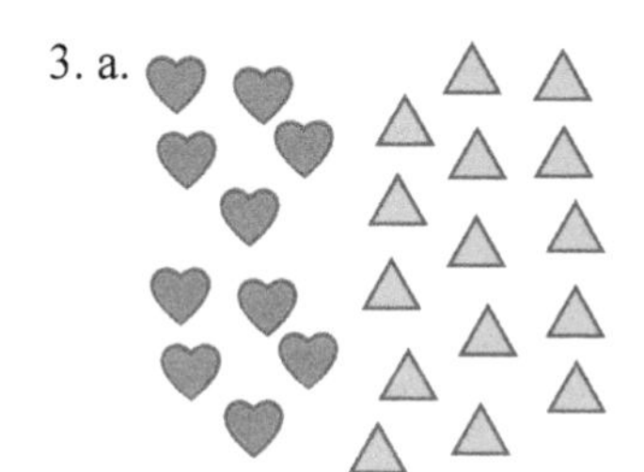

b. 2/3 heart for **1** triangle

3/2 triangles for **1** heart

4.

Miles	50	100	150	200	250
Hours	1	2	3	4	5

Miles	300	350	400	450	500
Hours	6	7	8	9	10

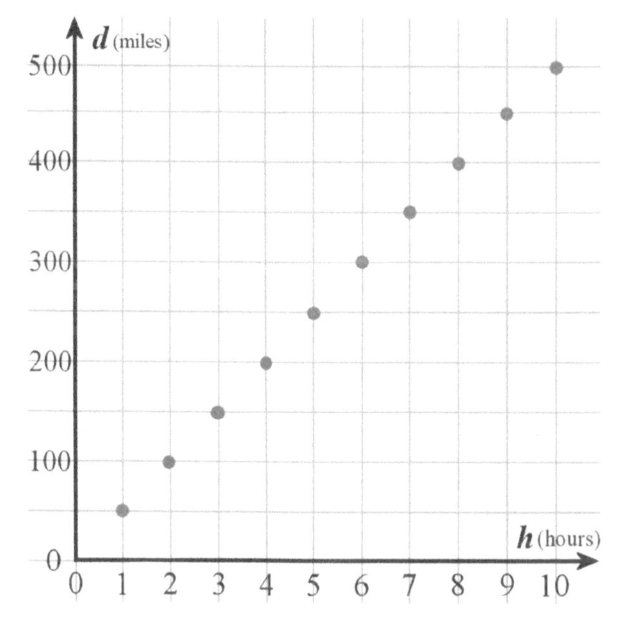

b. The unit rate is 50 miles per (1) hour.
c. The car would go 375 miles in 7 1/2 hours.
d. It would take 4 1/2 hours to travel 225 miles.

5. a. 20 g of salt : 1200 g of water = 1 : 60
   b. For 100 grams of salt you need 60 times as much water, which is 6,000 grams or 6 kg of water.

6. In the ratio 11:12, the 12 parts represent Dad's age (he's older) and the 11 parts represent Mom's. The 3-year difference in their ages is the one part difference in the ratio. Therefore, Dad is 3 × 12 = 36 years old and Mom is 33 years old.

7. A bean plant is 3/5 as tall as a tomato plant. The tomato plant is 20 cm taller than the bean plant.
   a. The ratio of the bean plant's height to the tomato plant's height is 3:5.
   b. The tomato plant is 20 cm or 2 parts taller than the bean plant so each part is 10 cm.
      The bean plant is 3 × 10 = 30 cm tall. The tomato plant is 5 × 10 = 50 cm tall.

8. 63 cm ÷ 9 × 16 = 112 cm. The television screen is 112 cm wide.

9. a. $20/12 kg × 5 kg = $8.33. Or, you can first find the unit rate. 20 dollars : 12 kg = 20/12 dollars per kg
      = 1 8/12 dollars per kg = 1 2/3 dollars per kg. Then, multiply that by 5 to get the price for 5 kg.
   b. The unit rate is 20 dollars : 12 kg = 20/12 dollars per kg = 1 8/12 dollars per kg = 1 2/3 dollars per kg = $1.67 per kg.

10.

a. $134 \text{ lb} = 134 \text{ lb} \cdot \frac{1 \text{ kg}}{2.2 \text{ lb}} = \frac{134 \text{ kg}}{2.2} \approx 60.9 \text{ kg}$
b. $156 \text{ cm} = 156 \text{ cm} \cdot \frac{1 \text{ in}}{2.54 \text{ cm}} \cdot \frac{1 \text{ ft}}{12 \text{ in}} = \frac{156 \text{ ft}}{2.54 \cdot 12} \approx 5.1 \text{ ft}$

Chapter 5: Percent

Percent, p. 170

1. a. shaded 67/100 = 0.67 = 67%; not shaded 33/100 = 0.33 = 33%
   b. shaded 4/100 = 0.04 = 4%; not shaded 96/100 = 0.96 = 96%

2.

a. 28% = $\frac{28}{100}$ = 0.28	b. 17% = $\frac{17}{100}$ = 0.17	c. 89% = $\frac{89}{100}$ = 0.89
d. 60% = $\frac{60}{100}$ = 0.60	e. 5% = $\frac{5}{100}$ = 0.05	f. 8% = $\frac{8}{100}$ = 0.08

3. a. About 7% of the babies have birth defects.
   b. About 93% percent of the babies do *not* have birth defects.
   c. You could expect to find about 35 babies with birth defects in a group of 500 babies.
   d. You could expect to find about 91 babies with birth defects in a group of 1,300 babies.

4.

a. $\frac{1}{2} = \frac{50}{100}$ = 50%	b. $\frac{1}{4} = \frac{25}{100}$ = 25%	c. $\frac{1}{5} = \frac{20}{100}$ = 20%

5.

a. $\frac{4}{10} = \frac{40}{100}$ = 40%	b. $\frac{11}{20} = \frac{55}{100}$ = 55%	c. $\frac{8}{10} = \frac{80}{100}$ = 80%
d. $\frac{3}{20} = \frac{15}{100}$ = 15%	e. $\frac{6}{25} = \frac{24}{100}$ = 24%	f. $\frac{4}{5} = \frac{80}{100}$ = 80%

6. a. shaded 3/5 = 60%; not shaded 2/5 = 40%
   b. shaded 3/4 = 75%; not shaded 1/4 = 25%
   c. shaded 8/10 = 80%; not shaded 2/10 = 20%

7.

a. $\frac{112}{100}$ = 1.12 = 112%	b. $\frac{109}{100}$ = 1.09 = 109%	c. $\frac{278}{100}$ = 2.78 = 278%

8.

a. 105% = $\frac{105}{100}$ = 1.05	b. 457% = $\frac{457}{100}$ = 4.57	c. 209% = $\frac{209}{100}$ = 2.09
d. 506% = $\frac{506}{100}$ = 5.06	e. 482% = $\frac{482}{100}$ = 4.82	f. 311% = $\frac{311}{100}$ = 3.11

9. a. About 4/5 (**80 %**) of the United States population is 14 years old or older.
   b. About 2/25 (**8 %**) of the world's population lives in North America.
   c. The continent of Africa covers about 1/5 (**20 %**) of the Earth's total land mass.

10. a. The taller one is 125% as tall as the smaller one.
    b. The smaller tree is 160 cm tall, so the taller tree is 200 cm tall.

11.

a. $\frac{3}{7}$ = 43%	b. $\frac{1}{8}$ = 13%	c. $\frac{5}{9}$ = 56%

Percent, cont.

12. a. About 1/20 (**5%**) of the population of India is 65 years old or older (2009 estimate).
    b. About 13/100 (**13%**) of the population of Australia is 65 years old or older (2009 estimate).
    c. The Indian Ocean covers approximately 7/50 (**14%**) of the Earth's surface.
    d. About 3/5 (**60%**) of the world's population lives in Asia.

13. a. 114%   b. 138%

Puzzle corner.
a. Each tiny white square is 1/64 of the whole. So the colored area as a fraction is (1/4) − (2/64) = 14/64 = 7/32.
   As a percentage, that is about 22%.

b. Each tiny colored triangle is 1/16 of the whole. So, the colored area is 9/16.
   As a percentage, this is about 56%.

c. Each tiny colored triangle is 1/16 of the whole. The colored area is 1/2 + 3/16 = 11/16.
   As a percentage, this is about 69%.

What Percentage . . . ?, p. 174

1. a. 20% (3/15 = 1/5 = 20/100)   b. 75% (12/16 = 3/4 = 75/100)

2. a. 75% (6/8 = 3/4 = 75/100)   b. 40% (120/300 = 12/30 = 2/5 = 40/100)   c. 80% (4/5 = 80/100)

3. a. 20% (2/10 = 20/100)   b. 62% (32/52 = 0.6154...)   c. 25% (24/96 = 1/4 = 25/100)

4.

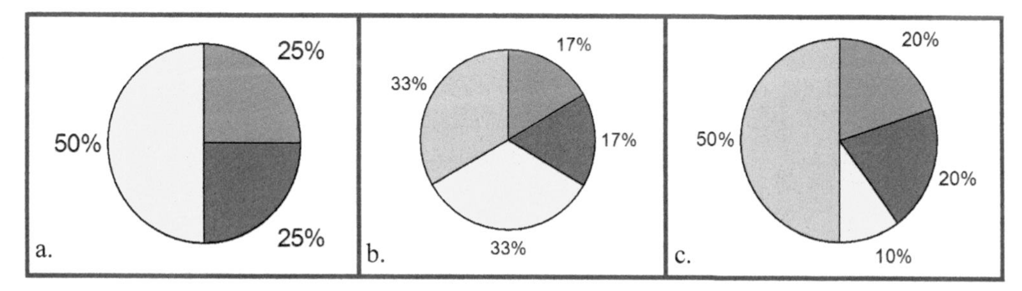

5. Divide each angle in degrees by 360.
   For example, 66° out of 360° is 66/360 = 0.18333... = 18%.

Step → ↓ Sector	Given: sector size (°)	Convert from degrees (÷ 360°)	Round to nearest whole %
1 (Gray)	15°	0.041666...	4%
2 (Brown)	32°	0.0888...	9%
3 (Yellow)	130°	0.36111...	36%
4 (Green)	66°	0.18333...	18%
5 (Violet)	117°	0.325	33%
TOTAL	360°	1 (circle)	100%

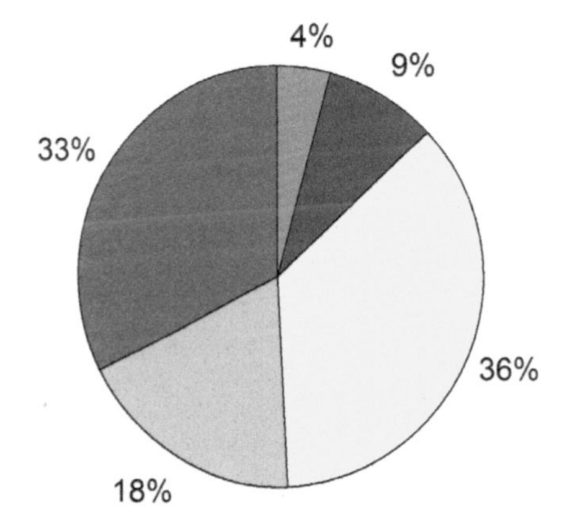

6. Of the 960 people at the conference, 450/960 = 0.468... ≈ 47% were doctors,
   220/960 = 0.229... ≈ 23% were nurses, and
   (960 − 450 − 220) / 960 = 290/960 = 0.302... ≈ 30% were researchers.

Puzzle Corner:
To produce the image at the right (→), find:
   47% of 360 degrees (169°),
   30% of 360 degrees (108°), and
   23% of 360 degrees (83°).
Then use those angles to draw the circle graph.

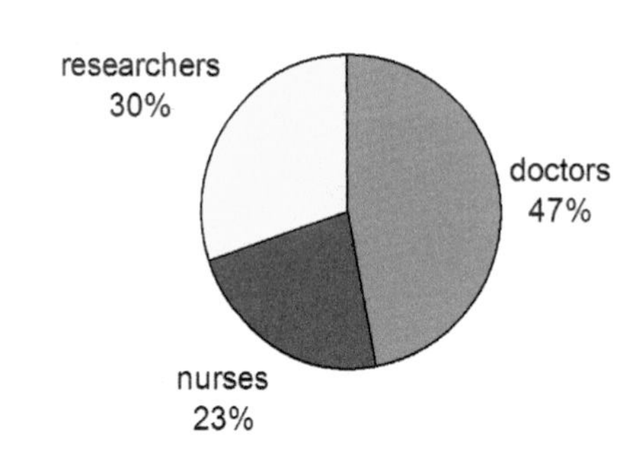

Percentage of a Number (Mental Math), p. 176

Can you think of a way to find 20% of a number? (Hint: Start with finding 10% of the number.)
Find 10% of the number and then double it.

1. a. 70 b. 32.1 c. 6 d. 0.7

2. a. 7 b. 3.21 c. 0.6 d. 0.07

3. Her total paycheck is $2,200.

4.

percentage / number	1,200	80	29	9	5.7
1% of the number	12	0.8	0.29	0.09	0.057
2% of the number	24	1.6	0.58	0.18	0.114
10% of the number	120	8	2.9	0.9	0.57
20% of the number	240	16	5.8	1.8	1.14

5.

Mental Math and Percent of a Number	
50% is $\frac{1}{2}$. To find 50% of a number, divide by **2**.	50% of 244 is **122**.
10% is $\frac{1}{10}$. To find 10% of a number, divide by **10**.	10% of 47 is **4.7**.
1% is $\frac{1}{100}$. To find 1% of a number, divide by **100**.	1% of 530 is **5.3**.
To find 20%, 30%, 40%, 60%, 70%, 80%, or 90% of a number, • First find **10%** of the number and • then multiply by 2, 3, 4, 6, 7, 8, or 9.	10% of 120 is **12**. 30 % of 120 is **36**. 60 % of 120 is **72**.

6.

a. 10% of 60 kg = 6 kg 20% of 60 kg = 12 kg	b. 10% of $14 = $1.40 30% of $14 = $4.20	c. 10% of 5 mi = 0.5 mi 40% of 5 mi = 2 mi
d. 1% of $60 = $0.60 4% of $60 = $2.40	e. 10% of 110 cm = 11 cm 70% of 110 cm = 77 cm	f. 1% of $1,330 = $13.30 3% of $1,330 = $39.90

7. a. The tax would be $420.
b. He would have $1,680 left after taxes.
c. He has 80% left.

8. Nancy has $2,170 left after paying taxes.

9. a. Peter figured if he subtracted 10% from the total amount, he would have 90% left. However, Peter subtracted 10% from the number 100 instead of from $55. The correct way to do it is to subtract $5.50 from $55. The correct answer is $49.50.

b. Patricia moved her decimal point one place too many to the left. One percent of $1,400 is actually $14. Multiply that by six to find the correct answer $84.

10.

a. 50% of 26 in = 13 in	b. 25% of 40 ft = 10 ft	c. 80% of 45 m = 36 m
d. 75% of $4.40 = $3.30	e. 90% of 1.2 m = 1.08 m	f. 25% of 120 lb = 30 lb

Percentage of a Number (Mental Math), cont.

11. 10% of \$65 is \$6.50. 1% of \$65 is \$0.65. 2% of \$65 is \$1.30.
    Now, add to get 12% of \$65: \$6.50 + \$1.30 = \$7.80

12. 25% of 44 kg is 11 kg. 1% of 44 kg is 0.44 kg.
    Subtract 11 kg − 0.44 kg = 10.56 kg

13. Hannah sent 51 messages during the night. She sent 289 messages during the day.

14. 100% − (30% + 45%) = 25% are white. Ten white horses is 25% of 40.

15. 945 students do not do either.
    To figure it out, first calculate 12% of 1,500, then calculate 25% of 1,500, and subtract both amounts from 1,500. Now, to find 12% of 1,500. First find 10% of 1,500, which is 150. 1% of 1,500 is 15, so 2% of 1,500 is double that or 30. Totaling the parts, we find that 12% of 1,500 is 150 + 30 = 180. To find 25% of 1,500, we can divide 1,500 ÷ 4 = 375. Then we subtract 1,500 − 180 − 375 = 945.

Percentage of a Number: Using Decimals, p. 179

1.

a. 20% of 70 0.2 × 70 = **14**	b. 90% of 50 0.9 × 50 = **45**	c. 80% of 400 0.8 × 400 = **320**
d. 60% of \$8 0.6 × \$8 = **\$4.80**	e. 9% of 3,000 0.09 × 3,000 = **270**	f. 7% of 40 L 0.07 × 40 = **2.8 L**
g. 150% of 44 kg 1.50 × 44 kg = **66 kg**	h. 200% of 56 students 2 × 56 = **112 students**	i. 2% of 1,500 km 0.02 × 1,500 km = **30 km**

2.

a. 0.6 × 50 60% of 50 = **30**	b. 0.03 × \$400 3% of \$400 = **\$12 s**	c. 0.8 × 400 mi 80% of 400 mi = **320 mi**
d. 0.08 × 6 8% of 6 = **0.48**	e. 0.11 × \$300 11% of \$300 = **\$33**	f. 0.2 × 70 kg 20% of 70 kg = **14 kg**

3. a. 0.17 × \$4500 = \$765
   b. 0.67 × 27 m = 18.09 m
   c. 0.48 × 7.8 kg = 3.744 kg

4. a. 0.25 × 240 mi = 60 mi
   b. 0.80 × 30,000 km = 24,000 km
   c. 0.75 × 3.2 kg = 2.4 kg

5. a. Since the whole shoreline is 100%, the percentage of the shoreline that is *not* sandy beach is 100% − 6% = 94%.
   b. 6% of 30 km means 0.06 × 30 km = 1.8 km of the shoreline is sandy beach.

6. a. 100% − 20% = 80% of the students do not have scholarships.
   b. 0.20 × 4,000 = 800 students have scholarships.
   c. 0.80 × 4,000 = 3,200 students do not have scholarships.

7. There were 900 acres planted in wheat, 1,350 acres in corn, and 750 acres in oats:

Crop	Percentage	Decimal	Acres Planted
Wheat	30%	0.30	900
Corn	45%	0.45	1350
Oats	100% − 30% − 45% = 25%	0.25	750
TOTAL	100%	1.00	3,000

Percentage of a Number: Using Decimals, cont.

8. a. Gladys did not convert 80% to a decimal. It should be 0.8 × 50 = 40.
   b. Glenn did not continue his solution far enough. He only found 25% of 84,000. After finding 25% of 84,000, he should have multiplied that by 3 to find 75% of 84,000. 84,000 ÷ 4 = 21,000, and 21,000 × 3 = 63,000.

9. Five of the expressions have the same value as 20% of $620:

~~0.02 × $620~~	$620 ÷ 5	$620 ÷ 10 × 2	2 × $62
$\frac{1}{5}$ × $620	0.2 × $620	~~20 × $620~~	~~$620 ÷ 4~~

10. About <u>27,470,000 more Argentines than Tanzanians live in cities</u>. First calculate 92% of 41 million Argentines: 0.92 × 41,000,000 = 37,720,000 Argentines. Then calculate 25% of 41 million Tanzanians: 0.25 × 41,000,000 = 10,250,000 Tanzanians. The difference is 37,720,000 − 10,250,000 = 27,470,000 more Argentines than Tanzanians. You can also solve it by first subtracting 92% − 25% = 67% and then multiplying 0.67 × 41,000,000.

11.

Activity	Percentage	Minutes	Hours/minutes
Sleep	38%	547	9 h 7 min
School	21%	302	5 h 2 min
Soccer	10%	144	2 h 24 min
Play	11%	158	2 h 38 min
Eating	9%	130	2 h 10 min
Chores	9%	130	2 h 10 min
Hygiene	2%	29	29 min
TOTAL	100%	1440	24 hours

b.

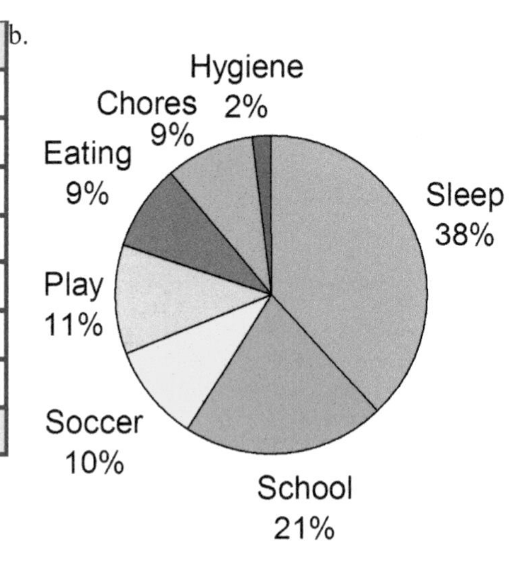

Discounts, p. 182

1. a. Sale price: $90 − $18 = <u>$72</u>.
   b. Discount amount: 0.40 × $5 = <u>$2</u>. Sale price: $5 − $2 = <u>$3</u>.
   c. Discount amount: 0.30 × $15 = <u>$4.50</u>. Sale price: $15 − $4.50 = <u>$10.50</u>.

2. Monica subtracted the percent discount as a dollar amount instead of calculating the amount of the discount in dollars. First she should have figured out 20% of $25, which is $5 (not $20). Then she should have subtracted $25 − $5 = $20. So the correct discounted price is $20.

3.

a. Discount amount: $0.30 Discounted price: $0.90	b. Discount amount: $4.50 Discounted price: $13.50	c. Discount amount: $45 Discounted price: $105
d. Discount amount: $8 Discounted price: $12	e. Discount amount: $0.22 Discounted price: $1.98	f. Discount amount: $0.65 Discounted price: $0.65

Discounts, cont.

4. a. Round \$39.90 ≈ \$40. Then calculate 30% of \$40 = \$12. The estimated discounted price is \$40 − \$12 = \$28.
   b. Round 17% to 20%. Then calculate 20% of \$12.50, which is \$2.50. The estimated discounted price is \$10.
   Or round 17% to 20% and \$12.50 to \$13. Then 20% of \$13 is 2.60, and the estimated discounted price is \$10.40.
   Or round 17% to 15% and \$12.50 to \$12. Then 15% of \$12 is \$1.80, and the estimated discounted price is \$10.20.
   Note: the exact answer is (1 − 0.17) × \$12.50 = \$10.38, so the second method, which calculates an estimated discounted price of \$10.40, is the most accurate. Why do you think that is? *
   c. Round \$75.50 to \$80. Then 75% of \$80 is \$60, and the estimated discounted price is \$20.

* Note to problem 4. b. ("Why do you think that is?"): By rounding the 17% discount to 20%, we are figuring ourselves a *bigger* discount than what we would really get. So by rounding the price *up* to a higher price, from \$12.50 to \$13.00, we are offsetting that error. When our errors of estimation cancel, our estimate is generally closer to the exact price.

5. a. Round \$199 to \$200. Since 75% of \$200 is \$150, the estimated discounted price would be \$50, which still costs more than the off-brand mp3 player for about \$45. So you'll need to decide if there is a big enough difference in quality between the two brands to justify paying \$5 more.
   b. Round \$89 to \$90. Then 40% of \$90 is \$36, and the estimated discounted price is about \$90 − \$36 = \$54, which is still more than about \$40 for a used copy. You will need to decide if the difference in condition is worth \$14 to you!

6. They would earn more money by selling it at the 25% discount. Without the discount, they are earning about 50 × \$40 = \$2,000 a week. Discounting it to around \$30 would make their estimated income about 100 × \$30 = \$3,000 a week.

7. a. The jeans were discounted by 10%. b. The phone was discounted by 25%. c. The haircut was discounted by 20%.

8. The expressions that are in gray will not work, but all of the others will.

0.25 × \$46 (gray)	0.75 × \$46	\$46 − \$46/25 (gray)	\$46 − \$46/4	\$46/4 (gray)	\$46/4 × 3

Practice with Percent p. 184

1. a. 0.10 × \$50 = \$5 b. \$10 / \$50 = 1/5 = 20%

2. a. Jenny ate 0.60 × 25 = 15 cookies. b. Jared ate 6/25 = 24/100 = 24% of the cookies.

3. a. Jack made baskets on 17/20 or 85% of his shots. b. Jack made 0.56 × 50 = 28 baskets in all.
   c. 0.60 × 25 = 15 of the women like chocolate. d. 42/200 = 21/100 = 21% of the citizens voted for Mr. X.
   e. 620/1000 = 62/100 = 62% of the boxes contained books. f. 0.14 × 50 = 7 of the participants came late.

4. a. 15/100 of the workers are over 50. (15 percent is a portion of 100 percent, not of 40 workers.)
   b. 6 workers are over 50. (15/100 × 40 workers = 6 workers. The 15 is a percent, not a quantity of workers.)

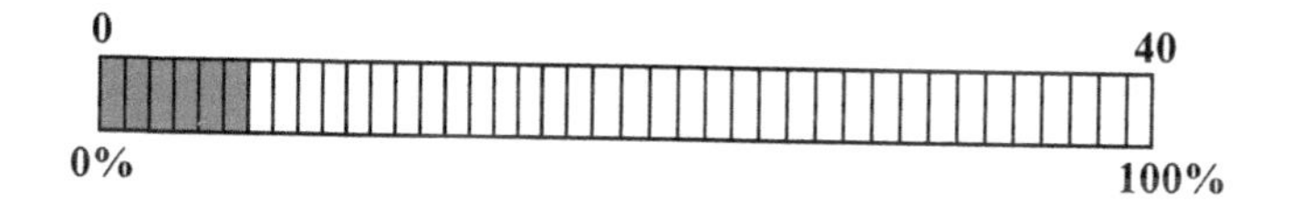

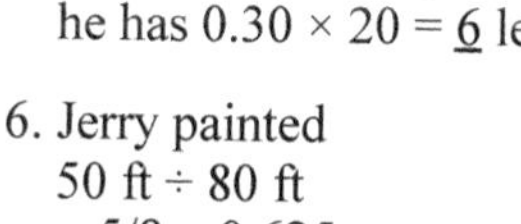

5. If Jose gave away 70% of his stuffed animals, then he kept 30%, so he has 0.30 × 20 = 6 left.

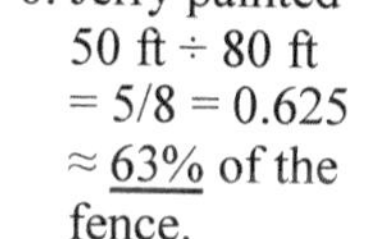

6. Jerry painted 50 ft ÷ 80 ft = 5/8 = 0.625 ≈ 63% of the fence.

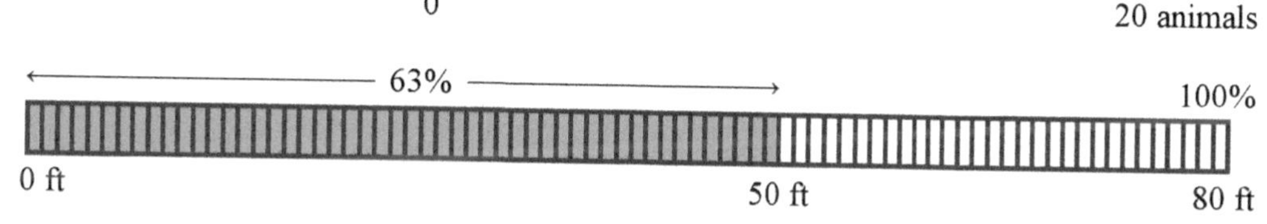

7. If Marie wasted 30% of her money, then she has 70% left. Since 0.70 × \$20 = \$14, she has \$14 left.

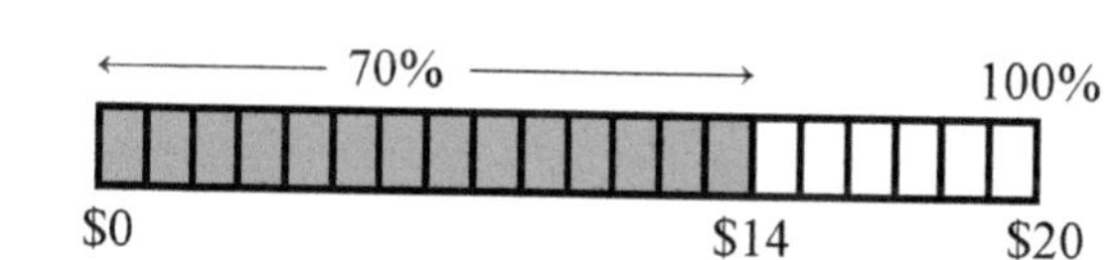

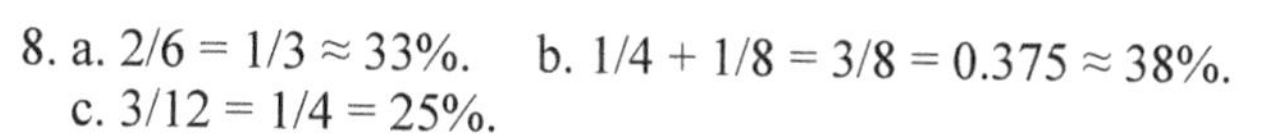

8. a. 2/6 = 1/3 ≈ 33%. b. 1/4 + 1/8 = 3/8 = 0.375 ≈ 38%.
   c. 3/12 = 1/4 = 25%.

9. a. Evelyn's height is 3 ft 9 in = (36 + 9) in = 45 in. Mary's height is 5 ft 2 in = (60 + 2) in = 62 in. Thus Evelyn's height is 45 in / 62 in = 0.7258... ≈ 73% of Mary's.
   b. Jacqueline is 90% × 6 ft 3 in = 0.9 × 75 in = 67.5 in = 5 ft 7 1/2 in tall.

Practice with Percent, cont.

10. a. Peter has to use a larger share of his wages because they pay the same, but he makes less.
    b. Each boy pays 50% of \$450, or \$225, a month for rent. So Peter pays \$225/\$900 = 25% of his wages, and Jake pays \$225/\$1,350 ≈ 17% of his wages.

11. a. 11% of \$402 is more. You can estimate that 11% of \$402 is about 10% of \$400, which is \$40. Similarly, 12% of \$298 is about 10% of \$300, which is \$30.
    b. Since 10% of \$400 is \$40 and 10% of \$300 is \$30, the difference is about \$10.

12. a. Their average mileage was the length of the total trip divided by how long it took them to travel it, or 1200 miles / 4 days = 300 miles per day.
    b. Day 1 340 miles / 1,200 miles = 0.2833... ≈ 28%
    Day 2 280 miles / 1,200 miles = 0.2333... ≈ 23%
    Day 3 400 miles / 1,200 miles = 0.3333... ≈ 33%
    Day 4 180 miles / 1,200 miles = 0.15 = 15%
    c. If they had divided the trip into four equal portions, on each of the four days they would have driven 1/4 = 25% of the total trip.

Finding the Total When the Percentage is Known, p. 187

1. Margie had 200 marbles. If 20% is 40 marbles, then 10% is 20 marbles. And 100% will be 10 times that, or 200 marbles.
2. Eric had \$600. If 15% is \$90, then 5% is \$30. And 100% is 20 times that, or \$600.
3. The ticket price was \$40 before the discount. The discounted price of \$16 is 40% of the original price. So 10% of the price is \$4. Therefore, 100% of the price is 10 times that, or \$40.
4. The dress cost \$30 before the discount. The discounted price of \$24 is 80% of the price. So 10% of the price is \$3, and 100% of the price is \$30.
5. One-half cup would be 25%. The recipe called for four times 1/2 cup of sugar, which originally was two cups of sugar.
6. Joe had \$200. What he has left (\$40) is 20% of his original money. So 10% of his money is \$20, and 100% of his money is \$200.
7. There were 80 people who had bought that brand of coffee grinder. The 72 people who were happy make up 90% of the total. So 10% is 72 ÷ 9 = 8, and 100% is ten times that, or 80 people.
8. Because both calculators have the same discounted price of \$42, the calculator that has the smaller discount (25%) was originally cheaper. \$42 is 70% of the price of the calculator that was discounted by 30%, so divide by 7 to find that 10% of the original price was \$6. The original price was 10 × \$6 = \$60. \$42 is 75% of the price of the calculator that was discounted by 25%, so divide \$42 by 3 to find that 25% of the original price was \$14, and multiply that amount by 4 to find the original price of \$56. Thus the calculator that was discounted by 25% was originally \$60 − \$56 = \$4 cheaper.

Puzzle corner: Joe had \$300 to begin with. The \$66 he has left is 100% − 78% = 22% of his original money. Now we can easily calculate 1% of the money: It is \$66 ÷ 22 = \$3, and 100% of the money is 100 times that, or \$300.

Mixed Review, p. 189

1. a. 13,054 R23; 13,054 × 26 + 23 = 339,427 b. 45 R69; 45 × 145 + 69 = 6,594
2. a. < b. > c. < d. < e. < f. <
3. a. 150 − 63 = 87 b. $\frac{8}{5} = 1\frac{3}{5} = 1.6$
4. $A = (3x \cdot 5x) + (2x \cdot 2x) = 15x^2 + 4x^2 = 19x^2$; $P = 5x + 3x + 5x + 1x + 2x + 2x + 2x = 20x$
5. a. $4y + 7$ b. $8r^3$
6. 30 ÷ 2.54 ≈ 11.8. Technically, the ruler is would be 11.8 inches long, but in reality, 30-cm rulers are made long enough to show 12 inches.

Mixed Review, cont.

7. The expressions that have the value of 6 are: a, c, d, f, g, i, j, and l.

8. a. The box weighs 280 grams. b. They would need to buy four boxes of paper clips.

9. $\dfrac{\$2}{5\text{ min}} = \dfrac{\$6}{15\text{ min}} = \dfrac{\$8}{20\text{ min}} = \dfrac{\$10}{25\text{ min}} = \dfrac{\$24}{1\text{ hr}}$

10. The sides of the rectangle make some multiple of the ratio 1:7, so let's call them x and $7x$. The perimeter is thus $x + 7x + x + 7x = 16x$. If the perimeter measures 120mm, then $x = 120\text{mm}/16 = 7.5$. So the width is $x = 7.5$mm, and the height is $7x = 52.5$mm.

11. Since 8:10 = 4:5 = 20:25, Gary can expect to make 20 baskets for every 25 shots.

12.

a.	b.	c.
$312 = x + 78$ $312 - 78 = x$ $234 = x$	$\frac{z}{2} = 60 + 80$ $\frac{z}{2} = 140$ $z = 2 \cdot 140$ $z = 280$	$7y - 2y = 45$ $5y = 45$ $y = 45 \div 5$ $y = 9$

13. $m = 0.3048 \cdot 89 \text{ ft} = 27.1272 \text{ m} \approx 27.13 \text{ m}$

14.

Distance	8	24 miles	40	48	120	144
Time	10 min	30 min	50 min	1 hour	2 1/2 hours	3 hours

Review: Percent, p. 191

1. a. 68% = $\frac{68}{100}$ = 0.68	b. 7% = $\frac{7}{100}$ = 0.07	c. 15% = $\frac{15}{100}$ = 0.15
d. 120% = $\frac{120}{100}$ = 1.20	e. 224% = $\frac{224}{100}$ = 2.24	f. 6% = $\frac{6}{100}$ = 0.06

2. percentage / number	6,100	90	57	6
1% of the number	61	0.9	0.57	0.06
4% of the number	244	3.6	2.28	0.24
10% of the number	610	9	5.7	0.6
30% of the number	1,830	27	17.1	1.8

3. There are 15 + 5 = 20 skaters. So 15/20 = 75% of the skaters are girls.

4. a. 75% b. 8% c. 163%

5. Emma's height is 133% of Madison's height. 64/48 = 133

6. The other chair costs \$35. 1.4 × 25 = 35

7. 25 = 1/5 of the marbles. 5 × 25 = 125 marbles total. 4 × 25 = 100 white marbles.

8. 540/2,000 = 54/200 = 27/100 = 0.27. Andrew pays 27% of his salary in taxes.

9. Since 0.80 × \$18 = \$14.40 and 0.90 × \$16 = \$14.40, they are both the same price.

10. The area of the square with 2-cm sides is 4 sq. cm., and the area of the square with 4-cm sides is 16 sq cm. Because 4/16 = 0.25, the area of the smaller square is 25% of the area of the larger square.

Math Mammoth Grade 6-B
Answer Key

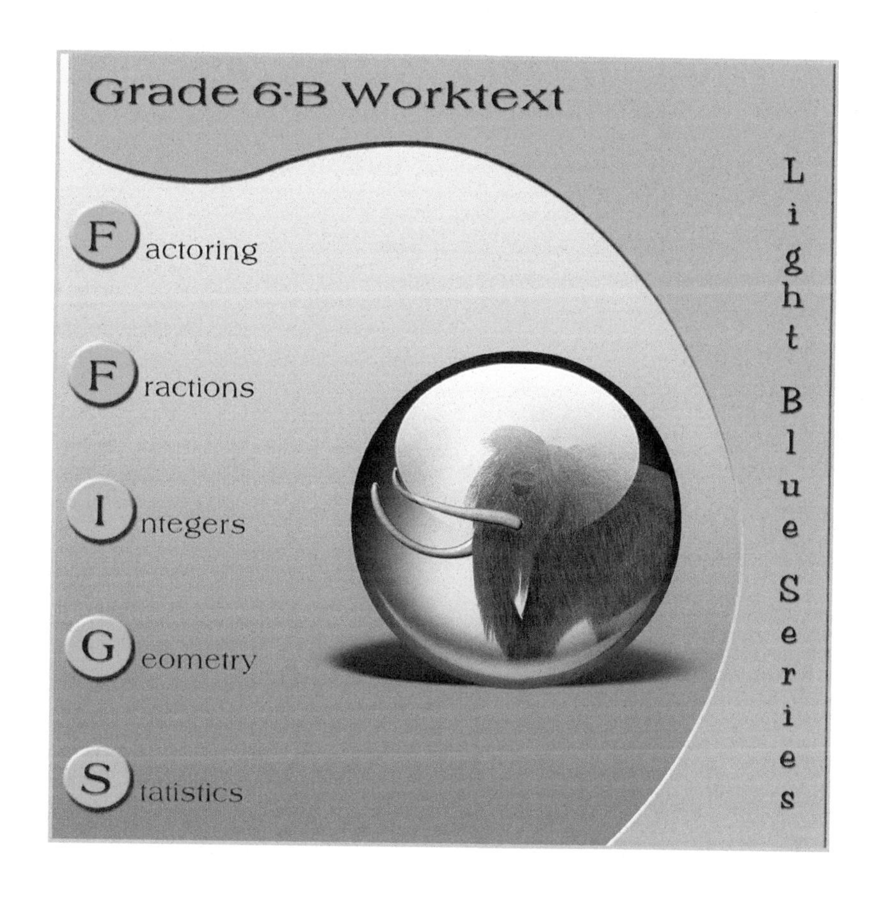

By Maria Miller

EDITION 12/2016

Math Mammoth Grade 6-B Answer Key
Contents

Chapter 6: Prime Factorization, GCF, and LCM

The Sieve of Eratosthenes and Prime Factorization, p. 10

List the primes between 0 and 100:
2, 3, 5, 7,
11, 13, 17, 19,
23, 29,
31, 37,
41, 43, 47,
53, 59,
61, 67,
71, 73, 79,
83, 89,
97

~~1~~	2	3	~~4~~	5	~~6~~	7	~~8~~	~~9~~	~~10~~
11	~~12~~	13	~~14~~	~~15~~	~~16~~	17	~~18~~	19	~~20~~
~~21~~	~~22~~	23	~~24~~	~~25~~	~~26~~	~~27~~	~~28~~	29	~~30~~
31	~~32~~	~~33~~	~~34~~	~~35~~	~~36~~	37	~~38~~	~~39~~	~~40~~
41	~~42~~	43	~~44~~	~~45~~	~~46~~	47	~~48~~	~~49~~	~~50~~
~~51~~	~~52~~	53	~~54~~	~~55~~	~~56~~	~~57~~	~~58~~	59	~~60~~
61	~~62~~	~~63~~	~~64~~	~~65~~	~~66~~	67	~~68~~	~~69~~	~~70~~
71	~~72~~	73	~~74~~	~~75~~	~~76~~	~~77~~	~~78~~	79	~~80~~
~~81~~	~~82~~	83	~~84~~	~~85~~	~~86~~	~~87~~	~~88~~	89	~~90~~
~~91~~	~~92~~	~~93~~	~~94~~	~~95~~	~~96~~	97	~~98~~	~~99~~	~~100~~

1.

a. $2^2 \times 31$	b. $2^2 \times 5 \times 13$	c. $2^5 \times 3$
d. $2 \times 3^2 \times 5$	e. $3 \times 5 \times 11$	f. 5×19
g. $2^4 \times 5$	h. $2^4 \times 3 \times 5$	i. $2^4 \times 17$
j. $2^2 \times 19$	k. $2 \times 3^2 \times 7$	l. $2^3 \times 13$

2.

a. $2^2 \times 7^2$	b. $2^2 \times 5 \times 19$	c. $2^4 \times 3 \times 7$
d. $2 \times 3^2 \times 17$	e. $2^2 \times 29$	f. $2^4 \times 3^2 \times 5$
g. $3^3 \times 5^2$	h. $2 \times 3^2 \times 5 \times 11$	i. $3^3 \times 5 \times 7$

Puzzle corner:
2, 3, 5, 7, 11, 13, 17, 19, 23, 29, 31, 37, 41, 43, 47, 53, 59, 61, 67, 71, 73, 79, 83, 89, 97, 101, 103, 107, 109, 113, 127, 131, 137, 139, 149, 151, 157, 163, 167, 173, 179, 181, 191, 193, 197, 199,

Using Factoring When Simplifying Fractions, p. 13

1.

a. $\frac{1}{3}$	b. $\frac{9}{11}$	c. $\frac{15}{23}$	d. $2\frac{1}{6}$
e. $\frac{5}{7}$	f. $1\frac{4}{15}$	g. $\frac{17}{24}$	h. $\frac{4}{5}$

2.

a. $\frac{19}{20}$	b. $\frac{33}{41}$	c. $\frac{23}{33}$
d. $\frac{1}{5}$	e. $\frac{19}{26}$	f. $\frac{6}{7}$

Using Factoring When Simplifying Fractions, cont.

3.

a. $\dfrac{\overset{7}{\cancel{14}}}{\underset{8}{\cancel{16}}} = \dfrac{7}{8}$	b. $\dfrac{\overset{11}{\cancel{33}}}{\underset{9}{\cancel{27}}} = \dfrac{11}{9} = 1\dfrac{2}{9}$	c. $\dfrac{\overset{6}{\cancel{12}}}{\underset{13}{\cancel{26}}} = \dfrac{6}{13}$	d. $\dfrac{\overset{3}{\cancel{9}}}{\underset{11}{\cancel{33}}} = \dfrac{3}{11}$	e. $\dfrac{\overset{6}{\cancel{42}}}{\underset{4}{\cancel{28}}} = 1\dfrac{1}{2}$

4. The exact manner of students' simplifications may vary.

a. $\dfrac{56}{84} = \dfrac{\overset{1}{\cancel{7}} \times \overset{2}{\cancel{8}}}{\underset{3}{\cancel{21}} \times \underset{1}{\cancel{4}}} = \dfrac{2}{3}$	b. $\dfrac{54}{144} = \dfrac{\overset{1}{\cancel{6}} \times \overset{3}{\cancel{9}}}{\underset{2}{\cancel{12}} \times \underset{4}{\cancel{12}}} = \dfrac{3}{8}$	c. $\dfrac{120}{72} = \dfrac{\overset{5}{\cancel{10}} \times \overset{4}{\cancel{12}}}{\underset{4}{\cancel{8}} \times \underset{3}{\cancel{9}}} = \dfrac{20}{12} = \dfrac{5}{3} = 1\dfrac{2}{3}$
d. $\dfrac{80}{48} = \dfrac{10 \times 8}{6 \times 8} = \dfrac{\overset{5}{\cancel{10}} \times \overset{1}{\cancel{8}}}{\underset{3}{\cancel{6}} \times \underset{1}{\cancel{8}}} = 1\dfrac{2}{3}$	e. $\dfrac{36}{90} = \dfrac{\overset{2}{\cancel{6}} \times \overset{\cancel{3}}{\cancel{6}}}{\underset{\cancel{3}}{\cancel{9}} \times \underset{5}{\cancel{10}}} = \dfrac{2}{5}$	f. $\dfrac{28}{140} = \dfrac{4 \times 7}{14 \times 10} = \dfrac{\overset{\cancel{2}}{\cancel{4}} \times \overset{1}{\cancel{7}}}{\underset{5}{\cancel{10}} \times \underset{\cancel{2}}{\cancel{14}}} = \dfrac{1}{5}$

5. The exact manner of students' simplifications may vary.

a. $\dfrac{14}{84} = \dfrac{\overset{1}{\cancel{2}} \times \overset{1}{\cancel{7}}}{\underset{3}{\cancel{21}} \times \underset{2}{\cancel{4}}} = \dfrac{1}{6}$	b. $\dfrac{54}{150} = \dfrac{\overset{3}{\cancel{9}} \times \overset{3}{\cancel{6}}}{\underset{5}{\cancel{15}} \times \underset{5}{\cancel{10}}} = \dfrac{9}{25}$	c. $\dfrac{138}{36} = \dfrac{\overset{1}{\cancel{2}} \times \overset{23}{\cancel{69}}}{\underset{2}{\cancel{4}} \times \underset{3}{\cancel{9}}} = \dfrac{23}{6} = 3\dfrac{5}{6}$
d. $\dfrac{\overset{9}{\cancel{27}}}{\underset{2}{\cancel{20}}} \times \dfrac{\overset{1}{\cancel{10}}}{\underset{7}{\cancel{21}}} = \dfrac{9}{14}$	e. $\dfrac{75}{90} = \dfrac{\overset{1}{\cancel{3}} \times \overset{5}{\cancel{25}}}{\underset{3}{\cancel{9}} \times \underset{2}{\cancel{10}}} = \dfrac{5}{6}$	f. $\dfrac{\overset{\overset{2}{\cancel{6}}}{\cancel{48}}}{\underset{\underset{3}{\cancel{9}}}{\cancel{45}}} \times \dfrac{\overset{11}{\cancel{55}}}{\underset{8}{\cancel{64}}} = \dfrac{\overset{1}{\cancel{2}}}{3} \times \dfrac{11}{\underset{4}{\cancel{8}}} = \dfrac{11}{12}$

6.

a. $\dfrac{22}{25}$	b. $\dfrac{14}{17}$	c. $\dfrac{17}{21}$
d. $\dfrac{16}{25}$	e. $\dfrac{3}{5}$	f. $\dfrac{13}{30}$
g. $\dfrac{5}{14}$	h. $\dfrac{17}{21}$	i. $\dfrac{9}{14}$

The Greatest Common Factor (GCF), p. 16

1. a. 1, 2, 3, 4, 6, 8, 12, 16, 24, 48
   b. 1, 2, 3, 4, 5, 6, 10, 12, 15, 20, 30, 60
   c. 1, 2, 3, 6, 7, 14, 21, 42
   d. 1, 3, 9, 11, 33, 99

2. a. 12 b. 6 c. 6 d. 3

3. a. 1, 2, 4, 11, 22, 44
   b. 1, 2, 3, 6, 11, 22, 33, 66
   c. 1, 2, 4, 7, 14, 28
   d. 1, 2, 4, 7, 8, 14, 28, 56
   e. 1, 2, 4, 5, 10, 20, 25, 50, 100
   f. 1, 3, 5, 9, 15, 45

4. a. 22 b. 4 c. 5 d. 3
   e. 4 f. 28 g. 4 h. 1

The Greatest Common Factor (GCF), cont.

5. a. $\frac{8}{11}$ b. $\frac{21}{22}$ c. $\frac{7}{8}$ d. $1\frac{13}{20}$

e. $\frac{12}{25}$ f. $1\frac{1}{99}$ g. 2 h. $\frac{4}{9}$

6. a. 6 b. 4 c. 2 d. 12 e. 1 f. 8

7. a. 5 b. 2 c. 6 d. 12

Puzzle corner:
Since $11 \times 17 = 187$ and $11 \times 24 = 264$, the greatest common factor of 187 and 264 is 11.

Factoring Sums, p. 19

1.

a. $3(x + 6)$ and $3 \cdot x + 3 \cdot 6 = 3x + 18$	b. $8(11 + b)$ and $8 \cdot 11 + 8 \cdot b = 88 + 8b$
c. $8(11 + 20)$ and $8 \cdot 11 + 8 \cdot 20 = 248$	d. $7(9 + 11)$ and $7 \cdot 9 + 7 \cdot 11 = 140$

2. 30 cm × 1 cm, 15 cm × 2 cm, 10 cm × 3 cm, and 6 cm × 5 cm

3. 40 cm × 1 cm, 20 cm × 2 cm, 10 cm × 4 cm, and 8 cm × 5 cm

4. Answers will vary. The answer here shows a length of 5 cm for the shared side.

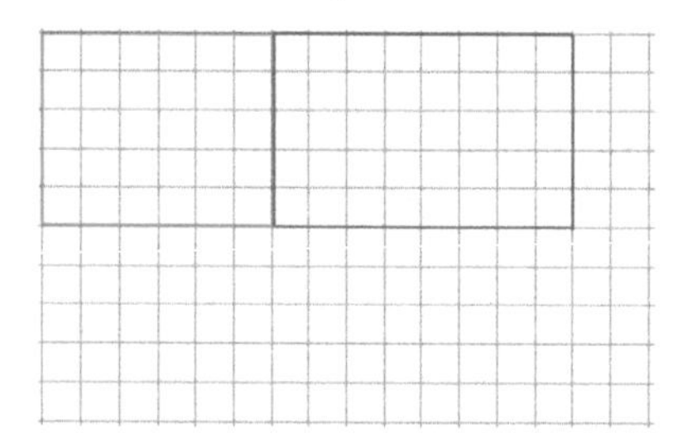

5. The sides of 1 cm, 2 cm, and 5 cm can all be used for the answer in problem 4. So, the rectangles that are side-by-side can be 30 cm × 1 cm and 40 cm × 1 cm, OR 15 cm × 2 cm and 20 cm × 2 cm, OR 6 cm × 5 cm and 8 cm × 5 cm.

6.

a. The GCF of 18 and 12 is 6. $18 + 12 = 6 \cdot 3 + 6 \cdot 2 = 6\,(3+2)$
b. The GCF of 6 and 10 is 2. $6 + 10 = 2 \cdot 3 + 2 \cdot 5 = 2(3 + 5)$
c. The GCF of 22 and 11 is 11. $22 + 11 = 11 \cdot 2 + 11 \cdot 1 = 11(2 + 1)$
d. The GCF of 15 and 21 is 3. $15 + 21 = 3 \cdot 5 + 3 \cdot 7 = 3(5 + 7)$
e. The GCF of 25 and 35 is 5. $25 + 35 = 5(5 + 7)$
f. The GCF of 72 and 86 is 2. $72 + 86 = 2\,(36 + 43)$
g. The GCF of 96 and 40 is 8. $96 + 40 = 8(12 + 5)$
h. The GCF of 39 and 81 is 3. $39 + 81 = 3(13 + 27)$

Factoring Sums cont.

7. a. $8(4 + 5)$

b.

8. The rectangles are 6×5 and 5×5.

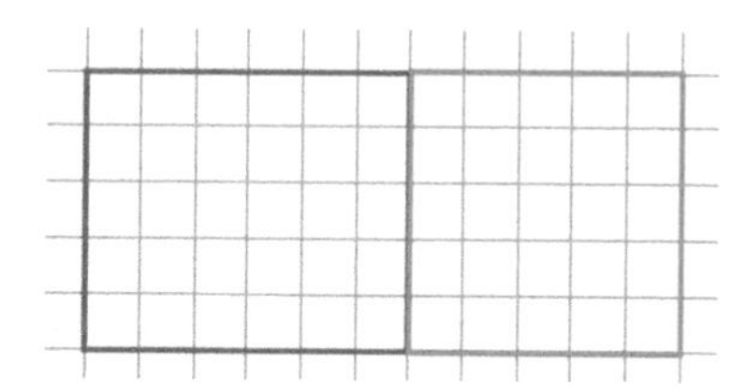

9. The rectangles are 7×6, 4×6, and 5×6.

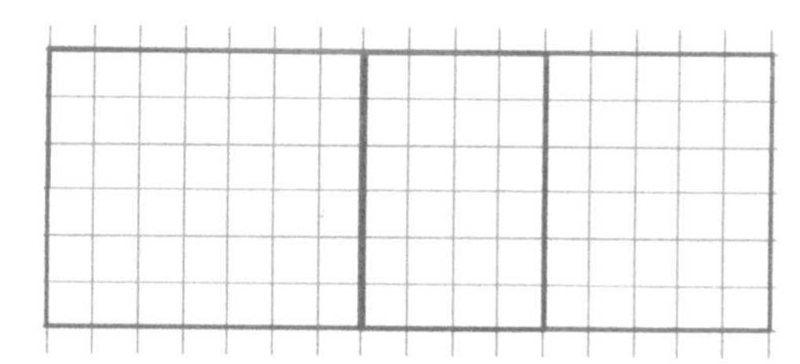

10. Since $45 = 3 \times 3 \times 5$, the pen can be 9 m × 5 m, 3 m × 15 m, or 45 m × 1 m. (The last one isn't a very likely size).

11. a. Answers will vary. Please check the student's answers. For example, 3 and 4, or 15 and 16. There is an infinite number of possible correct answers.

b. Answers will vary. Please check the student's answers. For example, 89 and 88, or 26 and 37.

12.

a. The GCF of 15 and 5 is 5. $15x + 5 = 5(3x + 1)$	b. The GCF of 18 and 30 is 6. $18x + 30 = 6(3x + 5)$
c. The GCF of 72 and 54 is 9. $72a + 54b = 9(8a + 6b)$	d. The GCF of 100 and 90 is 10. $100y + 90x = 10(10y + 9x)$

Puzzle corner:
a. Answers will vary. Please check the student's answers. Example: 26 and 39.
b. Answers will vary. Please check the student's answers. Example: 51 and 204.

The Least Common Multiple (LCM), p. 22

1. a. 6 b. 18 c. 56 d. 24 e. 70 f. 30

2. a. Answers will vary. Please check the student's answers.
   Any four of these numbers will work: 12, 18, 24, 30, 36, 42, 48, 54, 60, 66, 72, 78, 84, 90 or 96.

   b. Answers will vary. Please check the student's answers. For example: 1,250; 1,500; 1,750; 2,000.
   c. 96
   d. 1,100

3. a. Answers will vary. Please check the student's answers. Example: 30, 60, 90, 120.
   The LCM of 10 and 3 is 30.
   b. Answers will vary. Please check the student's answers. Example: 18, 36, 54, 72.
   The LCM of 6 and 9 is 18.
   c. Answers will vary. Please check the student's answers. Example: 24, 48, 72, 96.
   The LCM of 8 and 12 is 24.

4. a. no b. no c. no d. yes e. yes f. no

5. a. 12 b. 63 c. 10 d. 28 e. 10 f. 20

6. a. $\frac{17}{72}$ b. $\frac{23}{24}$ c. $\frac{9}{70}$ d. $\frac{13}{18}$

7. The greatest common factor is used in simplifying fractions, since we need to find a number that can divide ("go") into both the numerator and the denominator.
   The least common multiple is used in adding unlike fractions, since we need to find a number that both denominators can divide ("go") into.

8. a.

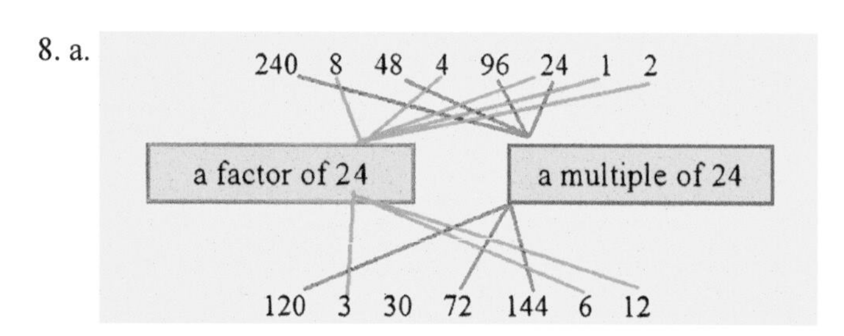

   b. The number 30 is neither a multiple nor a factor of 24.
   c. The number 24 is BOTH a factor and a multiple of 24.

9. a. 24 b. 30 c. 30 d. 56

10. a. 120 b. 120 c. 80 d. 600

11.

a. $\frac{29}{40}$ and $\frac{11}{15}$	b. $\frac{11}{20}$ and $\frac{13}{24}$	c. $\frac{7}{20}$ and $\frac{5}{16}$	d. $\frac{39}{50}$ and $\frac{94}{120}$
↓ ↓	↓ ↓	↓ ↓	↓ ↓
$\frac{87}{120} < \frac{88}{120}$	$\frac{66}{120} > \frac{65}{120}$	$\frac{28}{80} > \frac{25}{80}$	$\frac{468}{600} < \frac{470}{600}$

Puzzle corner: Day number 236 is Saturday. We need to divide 236 by seven, first of all, to find out how many complete weeks have passed. We also need take a careful look at the remainder to find how many additional days have passed.
236 ÷ 7 = 33 R5 So, 33 complete weeks and 5 additional days have passed.
The 33 completed weeks would take us to a certain Monday. 33 · 7 = 231 days. So, day 231 is a Monday. Then:
day 232 - Tuesday
day 233 - Wednesday
day 234 - Thursday
day 235 - Friday
day 236 - Saturday

Mixed Review, p. 26

1. a. 30,000 b. 343 c. 1,250

2. a. $109{,}200 = 1 \cdot 10^5 + 9 \cdot 10^3 + 2 \cdot 10^2$
   b. $7{,}002{,}050 = 7 \cdot 10^6 + 2 \cdot 10^3 + 5 \cdot 10^1$

3. The pieces are 3 ft 4.5 in (or 40.5 inches) and 5 ft 7.5 in (or 67.5 inches) long.
   Method: The easiest way is to work in inches, a smaller unit. The board is 9 × 12 = 108 inches long. The ratio of 3:5 means we're thinking of the board in 8 parts: 3 in the one piece and 5 in the other. Each part is 108 in ÷ 8 = 13.5 in. So the first piece is 3 × 13.5 in = 40.5 in, and the second piece is 5 × 13.5 in = 67.5 in.

4.

a. $\dfrac{16}{0.4} = \dfrac{160}{4} = 40$	b. $\dfrac{7}{0.007} = \dfrac{7{,}000}{7} = 1{,}000$	c. $\dfrac{99}{0.11} = \dfrac{9{,}900}{11} = 900$

5.

a. 100 × 0.2 = 20 120 × 0.02 = 2.4	b. 3 × 1.02 = 3.06 5 × 3.02 = 15.1	c. 0.9 × 0.2 × 0.5 = 0.09 30 × 0.005 × 0.2 = 0.03

6. a.

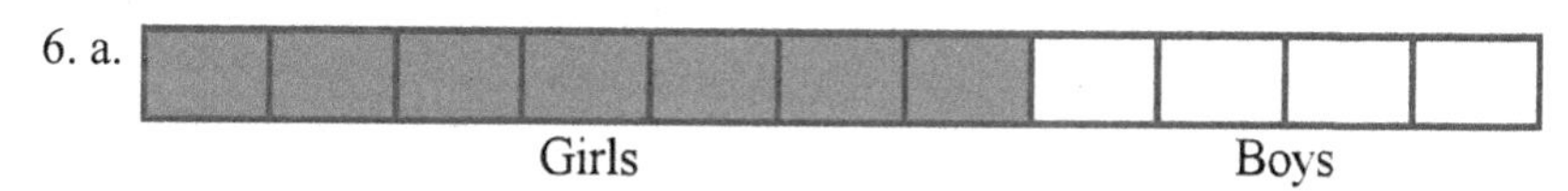

   b. The ratio of boys to all students is 4:11.
   c. There are 748 ÷ 11 × 7 = 476 girls.
      There are 748 ÷ 11 × 4 = 272 boys.

7. Liz's height is 150 cm/180 cm = 83.3% of her dad's height.

8. Their total budget for the month was \$540 ÷ 0.24 = \$2,250.

9. \$180 × 0.80 = \$144
   \$155 × 0.90 = \$139.50, which is cheaper.

10.

a. $2(7m + 4) = 2 \cdot 7m + 2 \cdot 4 = 14m + 8$	b. $10(x + 6 + 2y) = 10 \cdot x + 10 \cdot 6 + 10 \cdot 2y = 10x + 60 + 20y$

11. a. $\dfrac{5s + 8}{7}$ b. $(n + 11)^3$ c. $8 + y$ d. $\dfrac{x}{y^2}$

12. The numbers −2, −1, 0, 1, and 2 fulfill the inequality (any number less than 3 from the given set).

13. a. 3.093
    b. 0.206

Chapter 6 Review, p. 28

1.

a. 3 × 3 × 3 × 3	b. 2 × 13	c. 5 × 13
d. 3 × 2 × 2 × 2 × 2 × 2	e. 2 × 2 × 31	f. 2 × 3 × 3 × 5 × 5

2.

a. $\frac{28}{84} = \frac{\overset{1}{\cancel{4}} \times \overset{1}{\cancel{7}}}{\underset{3}{\cancel{21}} \times \underset{1}{\cancel{4}}} = \frac{1}{3}$	b. $\frac{75}{160} = \frac{\overset{1}{\cancel{5}} \times 15}{\underset{2}{\cancel{10}} \times 16} = \frac{15}{32}$
c. $\frac{222}{36} = \frac{\overset{1}{\cancel{6}} \times 37}{6 \times \underset{1}{\cancel{6}}} = \frac{37}{6} = 6\frac{1}{6}$	d. $\frac{48}{120} = \frac{\overset{1}{\cancel{6}} \times \overset{4}{\cancel{8}}}{\underset{2}{\cancel{12}} \times \underset{5}{\cancel{10}}} = \frac{4}{10} = \frac{2}{5}$

3. a. 21 b. 40 c. 66 d. 24

4. a. 8 b. 25 c. 16 d. 6

5. a. 25, 50, 75, 100, 125, and 150 are multiples of 25.
   1, 2, 5, 10, 25, and 50 are factors of 50.
   Each number has an infinite number of multiples.
   Each number has a greatest factor.
   If the number x divides into another number y, we say x is a factor of y.

   b. Answers will vary. Please check the student's answers.
   Any five of these will work: 75, 90, 105, 120, 135, 150, 165, 180, 195.

   c. Answers will vary. Please check the student's answers.
   Example: 28, 56, 112, 224, 448. The LCM of 4 and 7 is 28.

6.

a. GCF of 12 and 21 is 3. $12 + 21 = 3 \cdot 4 + 3 \cdot 7 = 3(4 + 7)$
b. GCF of 45 and 70 is 5. $45 + 70 = 5(9 + 14)$

7.

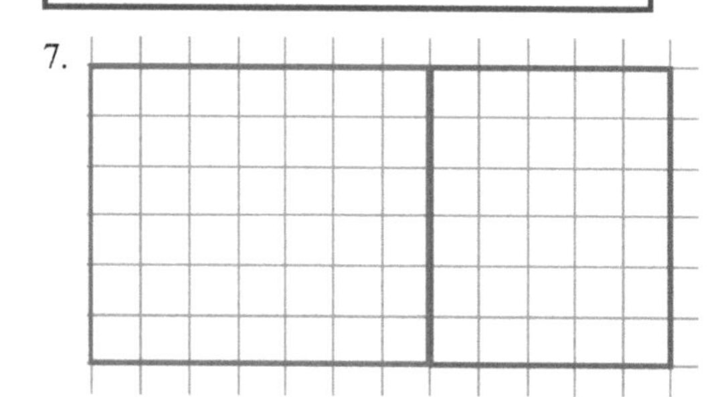

Chapter 7: Fractions

Review: Add and Subtract Fractions and Mixed Numbers, p. 34

1.

a.	b.	c.
$\frac{2}{5} + \frac{1}{2} \rightarrow = \frac{4}{10} + \frac{5}{10} = \frac{9}{10}$	$\frac{3}{4} + \frac{1}{9} \rightarrow \frac{27}{36} + \frac{4}{36} = \frac{31}{36}$	$\frac{7}{10} + \frac{1}{4} \rightarrow \frac{14}{20} + \frac{5}{20} = \frac{19}{20}$

2. a. 48 b. 36 c. 24 d. 60 e. 60 f. 100

3.

a. $\frac{15}{48} + \frac{8}{48} = \frac{23}{48}$	b. $3\frac{3}{36} + 1\frac{16}{36} = 4\frac{19}{36}$	c. $\frac{20}{24} - \frac{9}{24} = \frac{11}{24}$
d. $2\frac{25}{60} + \frac{48}{60} = 3\frac{13}{60}$	e. $5\frac{44}{60} - 2\frac{9}{60} = 3\frac{35}{60} = 3\frac{7}{12}$	f. $\frac{9}{20} + \frac{9}{20} = \frac{18}{20} = \frac{9}{10}$

4.

a.	b.	c.	d. / e.
$6\frac{12}{9} - 2\frac{7}{9} = 4\frac{5}{9}$	$17\frac{11}{10} - 5\frac{9}{10} = 12\frac{2}{10} = 12\frac{1}{5}$	$9\frac{16}{15} - 3\frac{8}{15} = 6\frac{8}{15}$	d. $15\frac{12}{9} - 9\frac{8}{9} = 6\frac{4}{9}$ e. $6\frac{17}{14} - 2\frac{10}{14} = 4\frac{7}{14} = 4\frac{1}{2}$

5.

a.	b.	c.
$3\frac{3}{4} \rightarrow 3\frac{9}{12}$ $-1\frac{1}{6} \rightarrow -1\frac{2}{12}$ $2\frac{7}{12}$	$3\frac{3}{8} \rightarrow 2\frac{33}{24}$ $-1\frac{5}{12} \rightarrow -1\frac{10}{24}$ $1\frac{23}{24}$	$8\frac{9}{11} \rightarrow 8\frac{18}{22}$ $-5\frac{1}{2} \rightarrow -5\frac{11}{22}$ $3\frac{7}{22}$

6.

Emma's way: $9\frac{2}{17} - 3\frac{8}{17}$

$= (9-3) + (\frac{2}{17} - \frac{8}{17}) = 6 - \frac{6}{17} = 5\frac{11}{17}$

Emma subtracted the whole numbers (9 – 3) and the fractions (2/17 – 8/17) separately.

Since the subtraction 2/17 – 8/17 actually "goes in the hole" by 6/17 or in other words yields a negative answer, she realized she needed to subtract 6/17 more from the whole numbers. So lastly she subtracted 6/17 from 6 to get the final answer.

Joe's method: $5\frac{3}{14} - 2\frac{9}{14}$

$\downarrow$

$5\frac{3}{14} - 2\frac{3}{14} - \frac{6}{14}$

$= 3 - \frac{6}{14} = 2\frac{8}{14}$

First, Joe subtracted what he could easily subtract without regrouping. So he subtracted 2 and 3/14. After that he had 3 left. Then he subtracted the rest (6/14).

Review: Add and Subtract Fractions and Mixed Numbers, cont.

7. Sarah used 5 3/4 cups of flour in the bread.

8. Joe's piece of wood is 9 in − 3 1/4 in = 5 3/4 inches long.

9.

a. $\frac{5}{12} + \frac{2}{12} + \frac{4}{12} = \frac{11}{12}$	b. $\frac{8}{28} + \frac{14}{28} - \frac{7}{28} = \frac{15}{28}$
c. $\frac{1}{10} + \frac{4}{10} = \frac{5}{10} = \frac{1}{2}$. Then, $\frac{1}{2} + \frac{2}{6} = \frac{5}{6}$	d. $\frac{57}{60} - \frac{20}{60} - \frac{15}{60} = \frac{22}{60} = \frac{11}{30}$
e. First, $\frac{35}{40} - \frac{8}{40} = \frac{27}{40}$. Then, $\frac{27}{40} + \frac{2}{3} = \frac{81}{120} + \frac{80}{120} = \frac{161}{120} = 1\,\frac{41}{120}$	f. $\frac{70}{60} - \frac{36}{60} + \frac{45}{60} = \frac{79}{60} = 1\,\frac{19}{60}$

10.

a. $\frac{1}{5} < \frac{2}{9}$	b. $\frac{15}{65} < \frac{15}{34}$	c. $\frac{6}{15} < \frac{1}{2}$	d. $\frac{3}{5} < \frac{8}{13}$
e. $\frac{1}{120} < \frac{1}{75}$	f. $\frac{2}{3} < \frac{8}{11}$	g. $\frac{11}{15} < \frac{3}{4}$	h. $\frac{10}{2000} > \frac{2}{1000}$

11.

a. $8\frac{1}{9} - 2\frac{5}{9} - 1 = 4\frac{5}{9}$	b. $5 - 2\frac{3}{24} - \frac{22}{24} = 1\frac{23}{24}$

12.

a. $x + \frac{1}{2} = 7\frac{1}{3}$ $x = 7\frac{1}{3} - \frac{1}{2}$ $x = 6\frac{5}{6}$	b. $x - 5\frac{7}{10} = 4\frac{3}{5}$ $x = 4\frac{3}{5} + 5\frac{7}{10}$ $x = 10\frac{3}{10}$
c. $8\frac{4}{7} + x = 10\frac{2}{5}$ $x = 10\frac{2}{5} - 8\frac{4}{7}$ $x = 1\frac{29}{35}$	d. $5\frac{1}{9} - x = 2\frac{1}{3}$ $x = 5\frac{1}{9} - 2\frac{1}{3}$ $x = 2\frac{7}{9}$

Add and Subtract Fractions: More Practice, p. 38

1\. a. 1 1/6 b. 19/20 c. 1 15/44 d. 5 11/20 e. 3 13/24 f. 12 59/100

2\.

a. $5\frac{1}{2} \rightarrow 5\frac{6}{12}$ $-\ 1\frac{7}{12} \rightarrow -\ 1\frac{7}{12}$ $3\frac{11}{12}$	b. $12\frac{1}{9} \rightarrow 12\frac{1}{9}$ $-\ 5\frac{2}{3} \rightarrow -\ 5\frac{6}{9}$ $6\frac{4}{9}$	c. $33\frac{1}{3} \rightarrow 33\frac{7}{21}$ $-\ 17\frac{6}{7} \rightarrow -\ 17\frac{18}{21}$ $15\frac{10}{21}$
d. $8\frac{1}{9} \rightarrow 8\frac{4}{36}$ $-\ 2\frac{7}{12} \rightarrow -\ 2\frac{21}{36}$ $5\frac{19}{36}$	e. $86\frac{6}{7} \rightarrow 86\frac{48}{56}$ $-\ 45\frac{1}{8} \rightarrow -\ 45\frac{7}{56}$ $41\frac{41}{56}$	f. $53\frac{1}{6} \rightarrow 53\frac{7}{42}$ $-\ 40\frac{6}{7} \rightarrow -\ 40\frac{36}{42}$ $12\frac{13}{42}$

3\.

I. $= 3\frac{4}{5}$

V. $= 11\frac{17}{30}$

U. $= 1\frac{23}{24}$

D. $= 1\frac{1}{4}$

E. $= 2\frac{3}{28}$

L. $= 3\frac{17}{36}$

I. $= 3\frac{31}{120}$

I. $= 4\frac{22}{27}$

N. $= \frac{7}{24}$

P. $= \frac{13}{35}$

I. $= 6\frac{3}{20}$

T. $= 6\frac{47}{60}$

G. $= 4\frac{7}{20}$

D. $= 2\frac{7}{22}$

M. $= 1\frac{11}{12}$

I. $= 1\frac{3}{35}$

D. $= 3\frac{1}{15}$

L. $= 3\frac{19}{24}$

Why did the amoeba flunk the math test? Because it...

$1\frac{11}{12}$	$1\frac{23}{24}$	$3\frac{17}{36}$	$6\frac{47}{60}$	$3\frac{4}{5}$	$\frac{13}{35}$	$3\frac{19}{24}$	$3\frac{31}{120}$	$2\frac{3}{28}$	$1\frac{1}{4}$
M	U	L	T	I	P	L	I	E	D

by

$2\frac{7}{22}$	$6\frac{3}{20}$	$11\frac{17}{30}$	$1\frac{3}{35}$	$3\frac{1}{15}$	$4\frac{22}{27}$	$\frac{7}{24}$	$4\frac{7}{20}$
D	I	V	I	D	I	N	G

Add and Subtract Fractions: More Practice, cont.

4. a. $3\frac{5}{15} - 1\frac{2}{15} + 1\frac{6}{15} = 3\frac{9}{15} = 3\frac{3}{5}$

b. $\frac{35}{50} + \frac{4}{50} + 1\frac{20}{50} = 2\frac{9}{50}$

c. $6\frac{67}{100} - 1\frac{40}{100} + 1\frac{55}{100} = 6\frac{41}{50}$

d. $2\frac{72}{48} - \frac{32}{48} - \frac{21}{48} = 2\frac{19}{48}$

e. $\frac{39}{48} + 2\frac{4}{48} + 2\frac{6}{48} = 5\frac{1}{48}$

f. $7\frac{7}{8} - 1\frac{4}{8} - 2\frac{2}{8} = 4\frac{1}{8}$

$5\frac{1}{48}$	$2\frac{7}{24}$	$2\frac{5}{48}$	$5\frac{3}{50}$	$1\frac{1}{60}$
$3\frac{11}{48}$	$6\frac{1}{8}$	$4\frac{5}{24}$	$3\frac{17}{24}$	$2\frac{1}{50}$
$3\frac{53}{66}$	$2\frac{19}{48}$	$2\frac{9}{50}$	$4\frac{1}{8}$	$3\frac{7}{8}$
$3\frac{1}{5}$	$3\frac{13}{15}$	$3\frac{3}{5}$	$5\frac{13}{66}$	$5\frac{4}{5}$
$4\frac{53}{66}$	$1\frac{29}{60}$	$4\frac{11}{50}$	$7\frac{39}{50}$	$6\frac{41}{50}$

g. $3\frac{21}{60} - 1\frac{5}{60} - 1\frac{15}{60} = 1\frac{1}{60}$

h. $5\frac{4}{24} + 1\frac{9}{24} - 2\frac{8}{24} = 4\frac{5}{24}$

i. $19\frac{42}{66} - 10\frac{22}{66} - 4\frac{33}{66} = 4\frac{53}{66}$

Review: Multiplying Fractions 1, p. 41

1.

a. $5 \times \frac{7}{8} = \frac{35}{8} = 4\frac{3}{8}$	b. $\frac{2}{7} \times \frac{5}{6} = \frac{10}{42} = \frac{5}{21}$
c. $\frac{9}{10} \times \frac{6}{7} \times \frac{1}{2} = \frac{54}{140} = \frac{27}{70}$	d. $\frac{4}{3} \times \frac{8}{3} = 3\frac{5}{9}$
e. $\frac{1}{10} \times \frac{16}{5} = \frac{8}{25}$	f. $\frac{17}{6} \times 10 \times \frac{1}{2} = \frac{170}{12} = 14\frac{1}{6}$

2. The volume of a cube with 1 1/4-inch sides is 1 61/64 cubic inches.

$$\frac{5}{4} \times \frac{5}{4} \times \frac{5}{4} = \frac{125}{64} = 1\frac{61}{64}$$

3. a. The area of one quilt square is 30 1/4 square inches.

$$\frac{11}{2} \times \frac{11}{2} = \frac{121}{4} = 30\frac{1}{4}$$

b. The total area of Mary's quilt is 5,445 square inches.

$$9 \times 20 \times \frac{121}{4} = \frac{21780}{4} = 5445$$

4.

E. $\frac{3}{10} \times \frac{1}{3} = \frac{1}{10}$

O. $\frac{2}{6} \times \frac{5}{7} = \frac{5}{21}$

M. $\frac{4}{10} \times \frac{1}{3} = \frac{2}{15}$

I. $7 \times \frac{5}{21} = \frac{5}{3}$

W. $\frac{4}{5} \times \frac{3}{6} = \frac{2}{5}$

S. $\frac{7}{40} \times 15 = \frac{21}{8}$

A. $\frac{5}{6} \times \frac{2}{4} = \frac{5}{12}$

L. $\frac{2}{9} \times \frac{9}{11} = \frac{2}{11}$

E. $\frac{3}{10} \times \frac{3}{9} = \frac{1}{10}$

N. $\frac{16}{24} \times 8 = \frac{16}{3}$

P. $\frac{4}{8} \times \frac{1}{3} = \frac{1}{6}$

R. $\frac{2}{6} \times \frac{3}{9} = \frac{1}{9}$

	$\frac{5}{12}$	$\frac{1}{9}$	$\frac{1}{10}$		$\frac{21}{8}$	$\frac{5}{3}$	$\frac{2}{15}$	$\frac{1}{6}$	$\frac{2}{11}$	$\frac{1}{10}$		$\frac{16}{3}$	$\frac{5}{21}$	$\frac{2}{5}$	
These problems	A	R	E		S	I	M	P	L	E		N	O	W	!

5.

a. Find $\frac{1}{2}$ of $\frac{1}{2} \times \frac{1}{4} = \frac{1}{8}$	b. Find $\frac{2}{3}$ of $\frac{2}{3} \times \frac{1}{2} = \frac{1}{3}$	c. Find $\frac{1}{4}$ of $\frac{1}{4} \times \frac{4}{5} = \frac{1}{5}$
d. Find $\frac{9}{10}$ of $\frac{9}{10} \times \frac{5}{6} = \frac{3}{4}$	e. Find $\frac{1}{6}$ of $\frac{1}{6} \times \frac{2}{3} = \frac{1}{9}$	f. Find $\frac{3}{8}$ of $\frac{3}{8} \times \frac{1}{2} = \frac{3}{16}$

6. Rewrite the ingredients for the pancake recipe as 3/4 of the original amounts.

Pancakes	Pancakes
1 3/4 c milk	1 5/16 milk
2 eggs	1 1/2 eggs
2 c flour	1 1/2 c flour
2 1/2 tsp baking powder	1 7/8 tsp baking powder
1/2 tsp salt	3/8 tsp salt
1 tsp cinnamon	3/4 tsp cinnamon

7. a. $\frac{3}{4} \times \frac{2}{3} \times 2 = \frac{2}{2}$ = 1 mile

b. 0.75 × 0.667 × 2 = 1.005 mile or about 1 mile. The reason we do not get the same answer as when using fractions is that 0.667 is not exactly 2/3 but is just an approximate decimal value for 2/3.

Review: Multiplying Fractions 2, p. 44

1.

a. $\frac{2}{3}\text{ m} \times \frac{3}{4}\text{ m} = \frac{6}{12} = \frac{1}{2}\text{ m}^2$	b. $\frac{2}{3}\text{ in} \times \frac{2}{3}\text{ in} = \frac{4}{9}\text{ in}^2$
c. $\frac{2}{3} \times \frac{2}{5} = \frac{4}{15}$	d. $\frac{3}{5} \times \frac{2}{3} = \frac{2}{5}$

2.

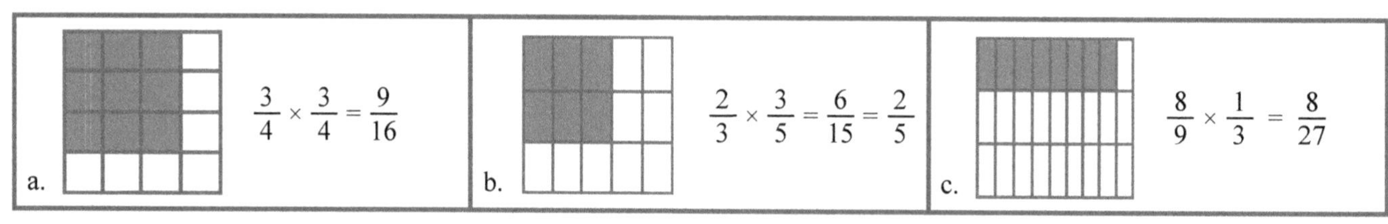

3.

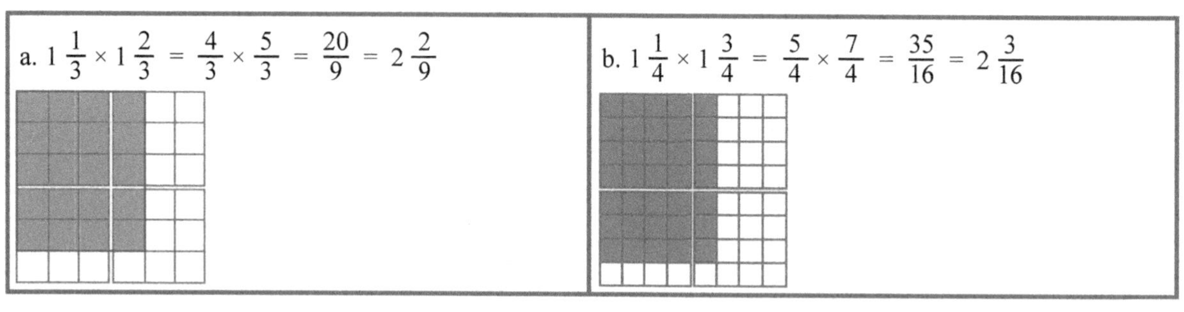

4. The area of the town is 14 11/32 square kilometers.

$$\frac{17}{4}\text{ km} \times \frac{27}{8}\text{ km} = \frac{459}{32}\text{ km}^2 = 14\frac{11}{32}\text{ km}^2$$

5. a. The area is 1 3/4 in^2.
The sides of the white rectangle are found by subtracting. The sides are
2 5/8 in − 7/8 in − 6/8 in = 1 in and 3 in − 7/8 in − 3/8 in = 1 6/8 in = 1 3/4 in.

So the area of the white rectangle is: $1\text{ in} \times 1\frac{3}{4}\text{ in} = 1\frac{3}{4}\text{ in}^2$

b. The area of the colored area is 6 1/8 square inches.

First, find the area of the whole rectangle: $3\text{ in} \times 2\frac{5}{8}\text{ in} = 3\text{ in} \times \frac{21}{8}\text{ in} = \frac{63}{8}\text{ in}^2 = 7\frac{7}{8}\text{ in}^2$.

Then, subtract from that the area of the white rectangle: $7\frac{7}{8}\text{ in}^2 - 1\frac{3}{4}\text{ in}^2 = 7\frac{7}{8}\text{ in}^2 - 1\frac{6}{8}\text{ in}^2 = 6\frac{1}{8}\text{ in}^2$

6. The price for 2 3/4 pounds of nuts is \$22.

$$\$8 \times 2\frac{3}{4} = \$16\frac{24}{4} = \$16 + \$6 = \$22$$

7. The price for 5/8 of a pound of nuts is \$6.25.

$$\$10 \times \frac{5}{8} = \frac{\$50}{8} = \$6\ 2/8 = \$6.25.$$

8. This time, change 1 3/8 into the decimal 1.375 and then multiply: \$10.38 × 1.375 = \$14.2725.
One and three-eighths pounds of nuts would cost \$14.27.

9.

a. $\frac{5}{5} \times \frac{2}{3} = \frac{10}{15} = \frac{2}{3}$	b. $\frac{3}{3} \times \frac{7}{10} = \frac{21}{30} = \frac{7}{10}$	c. $\frac{9}{9} \times \frac{8}{15} = \frac{72}{135} = \frac{8}{15}$

10.

a. $\frac{11}{12} \times 21 < 21$	b. $2\frac{1}{3} \times 19 > 19$	c. $\frac{16}{16} \times 105 = 105$

11.

a. $\frac{\overset{2}{\cancel{8}}}{\underset{2}{\cancel{12}}} \times \frac{\overset{1}{\cancel{6}}}{\underset{3}{\cancel{12}}} = \frac{2}{6} = \frac{1}{3}$	b. $\frac{\overset{1}{\cancel{3}}}{\underset{5}{\cancel{10}}} \times \frac{\overset{1}{\cancel{2}}}{\underset{6}{\cancel{18}}} = \frac{1}{30}$	c. $\frac{2}{\underset{3}{\cancel{30}}} \times \frac{\overset{1}{\cancel{10}}}{11} = \frac{2}{33}$
d. $\frac{7}{\underset{7}{\cancel{21}}} \times \frac{\overset{1}{\cancel{3}}}{4} = \frac{7}{28} = \frac{1}{4}$	e. $\frac{2}{\underset{2}{\cancel{16}}} \times \frac{\overset{1}{\cancel{8}}}{9} = \frac{2}{18} = \frac{1}{9}$	f. $\frac{\overset{2}{\cancel{18}}}{\underset{3}{\cancel{24}}} \times \frac{\overset{1}{\cancel{8}}}{\underset{1}{\cancel{9}}} = \frac{2}{3}$

12.

a. $\frac{\overset{1}{\cancel{5}}}{\underset{1}{\cancel{4}}} \times \frac{\overset{3}{\cancel{12}}}{\underset{3}{\cancel{9}}} \times \frac{\overset{1}{\cancel{3}}}{\underset{3}{\cancel{15}}} = \frac{3}{9} = \frac{1}{3}$	b. $\frac{\overset{1}{\cancel{8}}}{\underset{2}{\cancel{10}}} \times \frac{\overset{3}{\cancel{15}}}{\underset{3}{\cancel{27}}} \times \frac{\overset{1}{\cancel{9}}}{\underset{2}{\cancel{16}}} = \frac{3}{12} = \frac{1}{4}$
c. $\frac{1}{\underset{3}{\cancel{18}}} \times \frac{\overset{\overset{1}{\cancel{4}}}{\cancel{24}}}{\underset{11}{\cancel{33}}} \times \frac{\overset{3}{\cancel{9}}}{\underset{5}{\cancel{20}}} = \frac{3}{165} = \frac{1}{55}$	d. $\frac{\overset{1}{\cancel{3}}}{5} \times \frac{\overset{3}{\cancel{15}}}{\underset{\underset{3}{\cancel{9}}}{\cancel{18}}} \times \frac{\overset{8}{\cancel{16}}}{\underset{10}{\cancel{50}}} = \frac{24}{150} = \frac{4}{25}$

Puzzle corner

a. $\frac{\overset{6}{\cancel{60}}}{\underset{8}{\cancel{48}}} \times \frac{\overset{6}{\cancel{36}}}{\underset{9}{\cancel{90}}} = \frac{36}{72} = \frac{1}{2}$

b. $\frac{10}{3} \times \frac{5}{3} = \frac{50}{9} = 5\frac{5}{9}$

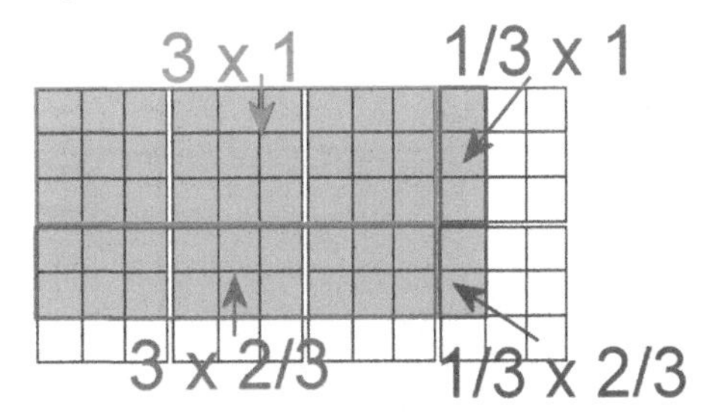

c. The method that your fellow student used is incomplete. The three complete squares in the upper left are the 3 × 1 = 3 part. The 2/9 of a square in the lower right is the 1/3 × 2/3 = 2/9 part.

He is missing two more parts. In the lower left we see that the 3 from the first factor also multiplies the 2/3 from the second factor: 3 × 2/3 = 2. Moreover, in the upper right we see that the 1/3 from the first factor also multiplies the 1 from the second factor: 1/3 × 1 = 1/3.

The solution requires the sum of all four of the factors: 3 + 2 + 1/3 + 2/9 = 5 5/9. Count the squares to verify that this is indeed the solution of the problem. This concept is the same as we use in algebra when we distribute multiplication over addition to solve expressions like $(x + 1) \cdot (x - 3)$.

Dividing Fractions: Reciprocal Numbers, p. 47

1.

a.  goes into 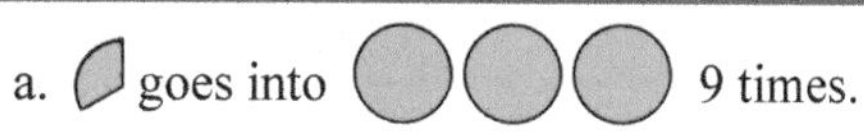 9 times. $3 \div \frac{1}{3} = 9$ Check: $9 \times \frac{1}{3} = 3$	b. goes into 7 times. $1\frac{3}{4} \div \frac{1}{4} = 7$ Check: $7 \times \frac{1}{4} = 1\frac{3}{4}$
c. 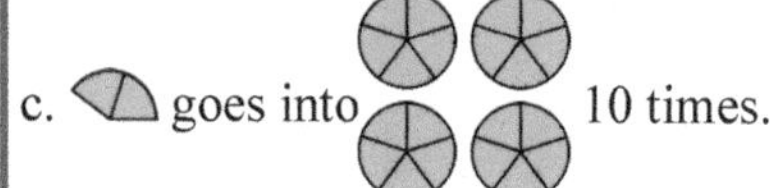 goes into 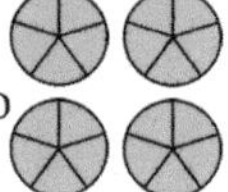 10 times. $4 \div \frac{2}{5} = 10$ Check: $10 \times \frac{2}{5} = \frac{20}{5} = 4$	d. goes into 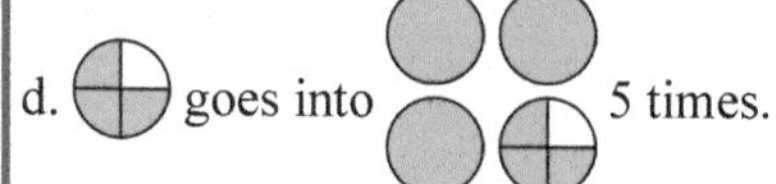 5 times. $3\frac{3}{4} \div \frac{3}{4} = 5$ Check: $5 \times \frac{3}{4} = \frac{15}{4} = 3\frac{3}{4}$

2. a. 18 b. 20 c. 30 d. 50 e. 28 f. 32 g. 40 h. 72

3.

a.	b.	c.	d.	e.
$\frac{8}{5}$	9	$\frac{8}{15}$	$\frac{1}{32}$	$\frac{8}{17}$
$\frac{5}{8} \times \frac{8}{5} = 1$	$\frac{1}{9} \times \frac{9}{1} = 1$	$\frac{15}{8} \times \frac{8}{15} = 1$	$32 \times \frac{1}{32} = 1$	$\frac{17}{8} \times \frac{8}{17} = 1$

4.

a. $1 \div \frac{5}{8} = \frac{8}{5}$	b. $1 \div \frac{1}{9} = \frac{9}{1} = 9$	c. $1 \div \frac{15}{8} = \frac{8}{15}$	d. $1 \div \frac{1}{32} = 32$	e. $1 \div \frac{17}{8} = \frac{8}{17}$

5.

a. 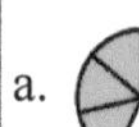 goes into 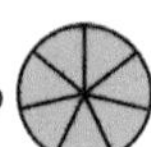 2 1/3 times. $1 \div \frac{3}{7} = \frac{7}{3} = 2\frac{1}{3}$	b. goes into 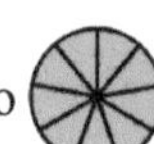 2 1/4 times. $1 \div \frac{4}{9} = \frac{9}{4} = 2\frac{1}{4}$
c. goes into 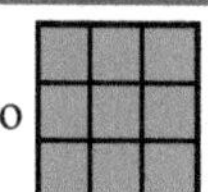 4 1/2 times. $1 \div \frac{2}{9} = \frac{9}{2} = 4\frac{1}{2}$	d. goes into 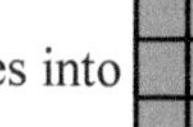 1 4/5 times. $1 \div \frac{5}{9} = \frac{9}{5} = 1\frac{4}{5}$
e. goes into 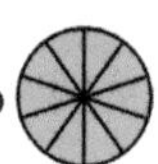 3 1/3 times. $1 \div \frac{3}{10} = \frac{10}{3} = 3\frac{1}{3}$	f.  goes into 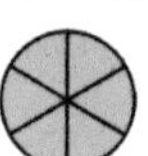 1 1/5 times. $1 \div \frac{5}{6} = \frac{6}{5} = 1\frac{1}{5}$

6.

a. 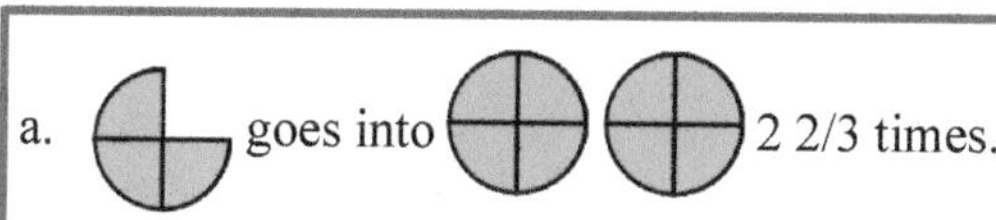

goes into 2 2/3 times.

$2 \div \frac{3}{4} = 2\frac{2}{3}$

b. goes into 4 times.

$1\frac{5}{7} \div \frac{3}{7} = \frac{12}{7} \div \frac{3}{7} = 4$

c. 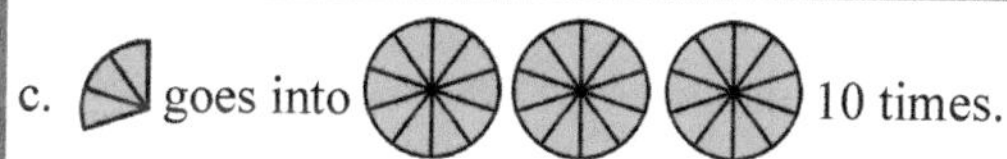

goes into 10 times.

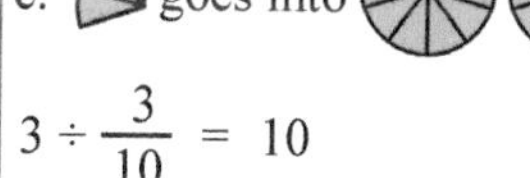

$3 \div \frac{3}{10} = 10$

d. 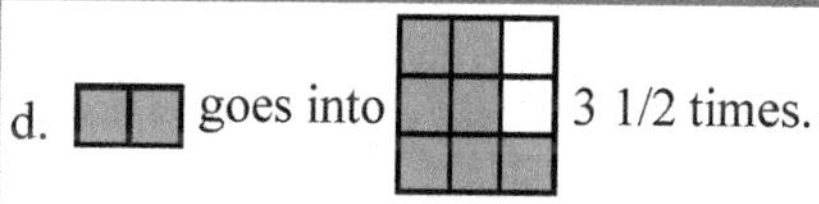

goes into 3 1/2 times.

$\frac{7}{9} \div \frac{2}{9} = 3\frac{1}{2}$

7.

a. $\frac{3}{4} \div 5$

↓ ↓

$\frac{3}{4} \times \frac{1}{5} = \frac{3}{20}$

b. $\frac{2}{3} \div \frac{6}{7}$

↓ ↓

$\frac{2}{3} \times \frac{7}{6} = \frac{7}{9}$

c. $\frac{4}{7} \div \frac{3}{7}$

↓ ↓

$\frac{4}{7} \times \frac{7}{3} = \frac{4}{3} = 1\frac{1}{3}$

d. $\frac{2}{3} \div \frac{3}{5}$

↓ ↓

$\frac{2}{3} \times \frac{5}{3} = \frac{10}{9} = 1\frac{1}{9}$

e. $4 \div \frac{2}{5}$

↓ ↓

$\frac{4}{1} \times \frac{5}{2} = 10$

f. $\frac{13}{3} \div \frac{1}{5}$

↓ ↓

$\frac{13}{3} \times \frac{5}{1} = \frac{65}{3} = 21\frac{2}{3}$

8. a. $1 \div \frac{8}{5} = \frac{5}{8}$

b. $\frac{5}{8} \times \frac{8}{5} = \frac{40}{40} = 1$

9.

$2 \div \frac{3}{4} = 2\frac{2}{3}$

Or, how many times does go into ?

First, let's solve how many times goes into .

Since $1 \div \frac{3}{4} = \frac{4}{3}$, it goes into one $\frac{4}{3} = 1\frac{1}{3}$ times.

If 3/4 fits into 4/3 = 1 1/3 times, then it fits into **double that many times**, or 8/3 = 2 2/3 times.

$2 \div \frac{3}{4}$

↓ ↓

$2 \times \frac{4}{3} = \frac{8}{3} = 2\frac{2}{3}$

Dividing Fractions: Reciprocal Numbers, cont.

10.

Or, how many times does 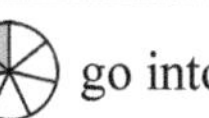 ? First, let's solve how many times 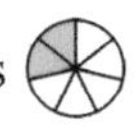 goes into . Since $1 \div \frac{2}{7} = \frac{7}{2}$, it goes into one 7/2 = 3 1/2 times. If 2/7 fits into 3 1/2 times, then it fits into exactly 5/6 as many times as it fits into 1, which is $\frac{5}{6} \times \frac{7}{2} = \frac{35}{12} = 2\frac{11}{12}$	$\frac{5}{6} \div \frac{2}{7}$ ↓ ↓ $\frac{5}{6} \times \frac{7}{2} = \frac{35}{12} = 2\frac{11}{12}$

Divide Fractions, p. 52

1.

a. $\frac{9}{10} \div \frac{2}{5}$ ↓ ↓ $\frac{9}{10} \times \frac{5}{2} = \frac{9}{4} = 2\frac{1}{4}$ Check: $\frac{9}{4} \times \frac{2}{5} = \frac{9}{10}$	b. $\frac{3}{7} \div \frac{4}{3}$ ↓ ↓ $\frac{3}{7} \times \frac{3}{4} = \frac{9}{28}$ Check: $\frac{9}{28} \times \frac{4}{3} = \frac{3}{7}$	c. $\frac{2}{11} \div \frac{2}{3}$ ↓ ↓ $\frac{2}{11} \times \frac{3}{2} = \frac{3}{11}$ Check: $\frac{3}{11} \times \frac{2}{3} = \frac{2}{11}$
d. $1\frac{7}{8} \div \frac{3}{4}$ ↓ ↓ $\frac{15}{8} \times \frac{4}{3} = \frac{5}{2} = 2\frac{1}{2}$ Check: $\frac{5}{2} \times \frac{3}{4} = 1\frac{7}{8}$	e. $2\frac{1}{15} \div 1\frac{3}{5}$ ↓ ↓ $\frac{31}{15} \times \frac{5}{8} = \frac{31}{24} = 1\frac{7}{24}$ Check: $\frac{31}{24} \times \frac{8}{5} = 2\frac{1}{15}$	f. $5\frac{10}{11} \div 6$ ↓ ↓ $\frac{65}{11} \times \frac{1}{6} = \frac{65}{66}$ Check: $\frac{65}{66} \times \frac{6}{1} = 5\frac{10}{11}$

2.

a. $\frac{8}{3} \div \frac{3}{4}$ ↓ ↓ $\frac{8}{3} \times \frac{4}{3} = \frac{32}{9} = 3\frac{5}{9}$ Check: $\frac{32}{9} \times \frac{3}{4} = \frac{8}{3}$	b. $\frac{7}{4} \div \frac{2}{3}$ ↓ ↓ $\frac{7}{4} \times \frac{3}{2} = \frac{21}{8} = 2\frac{5}{8}$ Check: $\frac{21}{8} \times \frac{2}{3} = \frac{7}{4}$

3.

a. Each person gets 5/6 of a pizza. $1\frac{4}{6} \div 2 = \frac{10}{6} \div 2 = \frac{5}{6}$	b. Each person gets 3/10 of a cake. $\frac{9}{10} \div 3 = \frac{3}{10}$

4.

a. $\frac{1}{2} \div 3 = \frac{1}{6}$	b. $\frac{1}{4} \div 3 = \frac{1}{12}$	c. $\frac{3}{5} \div 6 = \frac{6}{10} \div 6 = \frac{1}{10}$
d. $\frac{3}{4} \div 2 = \frac{3}{8}$	e. $\frac{1}{2} \div 5 = \frac{1}{10}$	f. $\frac{2}{3} \div 4 = \frac{4}{6} \div 4 = \frac{1}{6}$

5. 1/4 kg ÷ 3 = 1/12 kg

6. You can cut 29 pieces with 1/8 ft or 1.5 inches left over.
This can be solved in several ways. One way is division: 11 ft ÷ (3/8 ft) = 11 × (8/3) = 88/3 = 29 1/3.
So, you get 29 whole pieces and 1/3 of a piece. Since each piece is 3/8 ft long, the leftover 1/3 of a piece is 1/8 ft long.

Or, since 3/8 ft = 4 1/2 inches, and 11 ft = 132 inches, you can divide 132 by 4 1/2 and get the same result:
132 ÷ (4 1/2) = 132 ÷ (9/2) = 132 × (2/9) = 264/9 = 29 3/9 or 29 1/3. You get 29 pieces and 1/3 of a piece.

Or, you can guess and check and use multiplication: 30 × (3/8) ft = 90/8 ft = 11 2/8 ft. So, you do not get 30 pieces.
Instead, you get 29 pieces and 1/8 ft piece left over.

7. Each sibling inherited 12/25 of an acre. Divide 2 4/10 ÷ 5 = 24/10 ÷ 5 = (24/10) × (1/5) = 24/50 = 12/25.

8. a. 1/3 cup of wheat flour would make only 1/2 a batch. 1/3 ÷ 2/3 = 1/2.
b. 1 cup of wheat flour would make 1 1/2 batches. You can find that by dividing 1 cup ÷ (2/3 cup) = 1 × (3/2) = 1 1/2.

9. (2 1/8 mi) × (1/2 mi) = (17/8 mi) × (1/2 mi) = 17/16 sq. mi. = 1 1/16 sq. mi.

10. The width of the runway is 1/32 of a mile or 165 ft. You can either divide (1/16 sq. mi.) ÷ 2 mi = 1/32 mi, or you can think of multiplication: what number makes this sentence true: 2 mi × ______ = 1/16 sq. mi.

11.

a. $8x = \frac{1}{2}$ $x = \frac{1}{2} \div 8 = \frac{1}{16}$	b. $3x = \frac{3}{4}$ $x = \frac{3}{4} \div 3 = \frac{1}{4}$
c. $\frac{2}{3}x = \frac{1}{5}$ $x = \frac{1}{5} \div \frac{2}{3} = \frac{3}{10}$	d. $\frac{2}{3}x = 6$ $x = 6 \div \frac{2}{3} = 9$

12.

a. $\frac{1}{1} \div \frac{3}{4} = \frac{4}{3} = 1\frac{1}{3}$	b. $\frac{1}{1} \div \frac{3}{2} = \frac{2}{3}$	c. $\frac{1}{1} \div \frac{11}{7} = \frac{7}{11}$	d. $\frac{1}{1} \div \frac{9}{4} = \frac{4}{9}$

13. You can get 7 1/2 servings of 2/3 cup from 5 cups of ice cream. 5 ÷ (2/3) = 5 × (3/2) = 15/2 = 7 1/2.

14. The other sides measures 1/2 meter.
You can divide the area by the length of the one side to find the length of the other side: 2 1/2 m^2 ÷ 5 m = (5/2) ÷ 5 = (5/2) × (1/5) = 1/2 m, or you can think of multiplication: 5 m × ______ = 2 1/2 m^2.

15. a. The rectangle's sides are 3/8 in and 1/4 in.
The ratio of 2:3 means the two sides make up 5 "parts", and then the total perimeter consists of 10 parts.
Divide 1 1/4 in by 10 to find the length of one such part: 1 1/4 in ÷ 10 = 5/4 in ÷ 10 = 5/40 in.
Then, the two sides are 2 parts and 3 parts, or 10/40 in and 15/40 in, or 1/4 in and 3/8 in.
b. Check the student's drawing.

Divide Fractions, cont.

16. a. The area is 42 1/4 square feet. Multiply 6 1/2 ft × 6 1/2 ft = (13/2 ft) × (13/2 ft) = 169/4 sq. ft. = 42 1/4 sq. ft.
    b. The area is 6,084 sq. in. Each side is 78 inches, so the area is 78 in × 78 in = 6,084 sq. in.
    c. One quarter of the garden would be 10 9/16 square feet. You find the area by dividing 42 1/4 sq. ft ÷ 4.
    One way to do this division is to first divide 40 sq. ft by 4, then 2 sq. ft by 4, and lastly 1/4 sq. ft by 4, and add the results. 40 sq. ft. ÷ 4 = 10 sq. ft., 2 sq. ft. ÷ 4 = 1/2 sq. ft., and 1/4 sq. ft. ÷ 4 = 1/16 sq. ft. In total, one-fourth of the garden has an area of 10 9/16 sq. ft.

Problem Solving with Fractions 1, p. 56

1. The printable area would be 7 1/2 by 10 inches, which has an area of 75 square inches (7.5 in × 10 in = 75 sq. in.).
2. a. She needs to make the recipe 7 1/2 times (30 ÷ 4 = 7 1/2).

a.
Spiced Coffee – 30 servings
11 1/4 teaspoons of ground cinnamon
3 3/4 teaspoons of ground nutmeg
15 tablespoons of sugar
7 1/2 cups of heavy cream
22 1/2 cups of coffee
30 teaspoons of chocolate syrup

b.
Spiced Coffee – 1 serving
3/8 of a teaspoon of ground cinnamon
1/8 of a teaspoon of ground nutmeg
1/2 of a tablespoon of sugar
1/4 cup of heavy cream
3/4 cup of coffee
1 teaspoon of chocolate syrup

3. 1/5 of $45.55 is $9.11, and taking that off the price leaves $36.44 to pay.
   1/4 of $52.80 is $13.20, and taking that off the price leaves $39.60 to pay.
   So, the book that is 1/5 off is cheaper (or is the better deal).
4. a. Each "1/2 of all" is 472 ÷ 2 = 236. Then 1/4 of that is 236 ÷ 4 = 59. The unknown, x, is 3 × 59 = 177.
   b. One-third of 930 is 930 ÷ 3 = 310. Two-thirds of 930 is double, or 620. Then, to find x, we find 2/5 of that 620.
   One-fifth of 620 is 124. Two-fifths of 620 is double that, or 248.

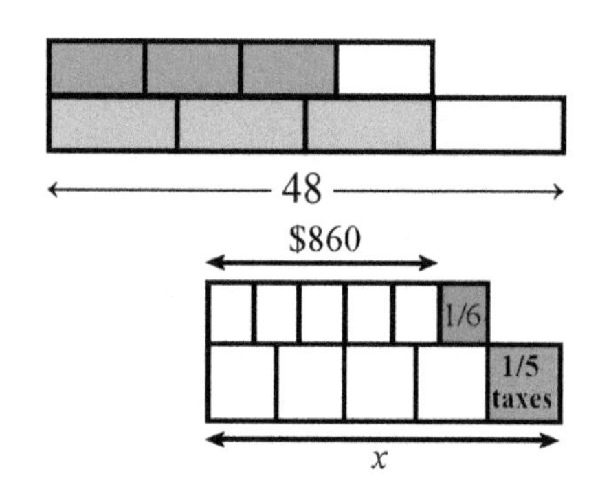

5. Three-fourths of 48 flowers is 36 flowers, so Elaine gave 36 flowers to her grandmother. Her grandmother gave 1/4 of those, or 9 flowers, to her neighbor. So her grandmother has 27 flowers now. The neighbor got 9 flowers. And Elaine has 12 flowers left.
6. Dad's paycheck was $1,290. See the bar model on the right. First, find how much one block in the top bar is, and multiply that by 6 to get the total amount that was left after taxes (the entire top bar). We get $860 ÷ 5 × 6 = $1,032. This amount ($1,032) is the entire top bar and at the same time is 4/5 of the paycheck. Now divide that amount by 4 (to get one block in the lower bar) and then multiply the result by 5 to get the original amount of the paycheck: $1,032 ÷ 4 × 5 = $1,290.
7. Dad has $600 left now. Dad paid $200 for car repairs and had $1000 left.
   So the grocery bill was 2/5 of $1000, or $400.
8. a. Together, the farmer and the wholesale dealer get 1/8 + 1/12 = 3/24 + 2/24 = 5/24.
   The merchant gets 19/24 of the money from the potato sales.
   b. The farmer would get $562.50. The wholesale dealer would get $375. The store keeper would get $3,562.50.
9.

a.	b.	c.	d.
$3 \div \frac{1}{5} = 15$	$6 \div \frac{1}{4} = 24$	$1 \div \frac{1}{4} = 4$	$8 \div \frac{1}{2} = 16$
$3 \div \frac{2}{5} = 7\frac{1}{2}$	$6 \div \frac{2}{4} = 12$	$2 \div \frac{1}{4} = 8$	$8 \div \frac{2}{2} = 8$
$3 \div \frac{3}{5} = 5$	$6 \div \frac{3}{4} = 8$	$3 \div \frac{1}{4} = 12$	$8 \div \frac{3}{2} = 5\frac{1}{3}$
$3 \div \frac{4}{5} = 3\frac{3}{4}$	$6 \div \frac{4}{4} = 6$	$4 \div \frac{1}{4} = 16$	$8 \div \frac{4}{2} = 4$
$3 \div \frac{5}{5} = 3$	$6 \div \frac{5}{4} = 4\frac{4}{5}$	$5 \div \frac{1}{4} = 20$	$8 \div \frac{5}{2} = 3\frac{1}{5}$

Problem Solving with Fractions 2, p. 59

1. There are a total of eight parts. 52 oz ÷ 8 = 6.5 oz. There were 6.5 oz of juice concentrate and 45.5 oz of water.

2. The sides are 7/8 in and 4 3/8 in. The ratio 1:5 means that there are a total of six "parts" for the two sides, or 12 parts for the perimeter. To find the length of one part, divide 10 1/2 in by 12: 10 1/2 in ÷ 12 = 21/2 in ÷ 12 = 21/24 in = 7/8 inch. Then, the width is one of those parts, or 7/8 inch, and the length is five of those parts, or 4 3/8 inches.

3. The longer route is 1 1/2 miles long.
   Divide the shorter route's length by 3 to get one "part": 1 1/8 mi ÷ 3 = (9/8 mi) × 1/3 = 9/24 mi = 3/8 mile.
   Multiply that distance by 4 to get the length of the longer route: 4 × (3/8 mi) = 12/8 mi = 1 1/2 mi.

4.

a. 3/4 of a number is 15. 1/4 of that number is 5. The number is 20.	b. 2/9 of a number is 24. 1/9 of that number is 12. The number is 108.	c. 7/8 of a number is 49. 1/8 of that number is 7. The number is 56.

5. a. 30 b. 60 c. 77 d. 63

6. The club has 36 members. The 8 boys are 2/9 of the total, so 1/9 of the total is 4. Therefore, the total is 9 times that, or 36.

7. 14 more gallons of gas. Divide 21 gal ÷ 3 = 7 gal to find how many gallons would plow 1/5 of the farm. Then, since the rest of the farm is 2/5 of the total, 2 × 7 gallons = 14 gallons gives you how many gallons are needed for the rest.

8. The total income was $13,413.75.
   There were 657 ÷ 3 = 219 seniors and 657 – 219 = 438 non-seniors. The income from seniors' tickets was 219 × $12.25 = $2,682.75, and the income from the others tickets was 438 × $24.50 = $10,731. The total income was $10,731 + $2,682.75 = $13,413.75.

9. a. 32 erasers
   b. 4 erasers. Divide 6 in ÷ (1 3/8 in) = 6 ÷ (11/8) = 48/11 = 4 4/11. Since we are not fitting *parts* of erasers into the box, the answer is simply 4 (ignore the 4/11 of an eraser in the answer).
   c. 6 erasers. Divide 5 in ÷ (13/16 in) = 80/13 = 6 2/13. As above, we ignore the 2/13 part of an eraser.
   d. 32 × 4 × 6 = 768 erasers would fit into the box.

10. Way 1:
    Length: 12 in ÷ (11/8 in) = 96/11 = 8 8/11.
    Width: 10 in ÷ (13/16 in) = 160/13 = 12 4/13.
    Height: 8 in ÷ (1/8 in) = 64.

    So 8 × 12 × 64 = 6,144 erasers would fit in the box.

    Way 2: If we change the orientation of the erasers, and put the 13/16 in sides along the 12-inch side of the box:

    Length: 12 in ÷ (13/16 in) = 192/13 = 14 10/13.
    Width: 10 in ÷ (11/8 in) = 80/11 = 7 3/11.
    Height: 8 in ÷ (1/8 in) = 64.

    So, 14 × 7 × 64 = 6,272 erasers would fit in the box.

    There are several more ways to orient the erasers. For example, they could be placed standing up.

11. It was 136 cm. Half of the piece was 17 × 4 = 68 cm. The whole piece was 68 cm × 2 = 136 cm.

12. He harvested 93 kg of apples. The 15.5 kg is 1/6 of the original harvest. The total was therefore 15.5 kg × 6 = 93 kg.

Mixed Review, p. 62

1. a. There are 12 inches in one foot, so in one mile we have about 5,000 × 12 = 60,000 inches. Or you can estimate it as 5,300 × 12 = 63,600 inches
   b. There are exactly 63,360 inches in one mile.

2. There are 16 ounces of juice concentrate and 48 ounces of water. There are a total of 8 "parts". Each part is 64 oz ÷ 8 = 8 oz. So, there are 2 × 8 oz = 16 oz of juice concentrate and 6 × 8 oz = 48 oz of water.

3.

a.	b.
$\frac{\$80}{4\text{ hr}} = \frac{\$20}{1\text{ hr}} = \frac{\$60}{3\text{ hr}} = \frac{\$5}{15\text{ min}}$	$\frac{2\text{ m}^2}{5\text{ min}} = \frac{10\text{ m}^2}{25\text{ min}} = \frac{120\text{ m}^2}{5\text{ hours}} = \frac{250\text{ m}^2}{10\text{ hr }25\text{ min}}$

4. There are 24 grams of salt and 1,176 grams of water.
   The mixture weighs 1,200 g. So, 1% of it is 12 g and 2% of it is 24 g.

5. The train left at 12:50 p.m.
   The table below helps you find how much time it took the train to travel 165 miles:

3 miles	15 miles	45 miles	90 miles	135 miles	150 miles	165 miles	180 miles
2 min	10 min	30 min	1 h	1 1/2 h	1 h 40 min	1 h 50 min	2 h

   Going 1 h 50 min backwards from 2:40 p.m. gives us 12:50 p.m.

6.

a.	b.	c.
10 × 0.3909 = 3.909 1,000 × 4.507 = 4507	1.08 × 100 = 108 $0.0034 \times 10^4 = 34$	$10^6 \times 8.02 = 8020000$ $10^5 \times 0.004726 = 472.6$
d.	**e.**	**f.**
0.93 ÷ 100 = 0.0093 48 ÷ 10 = 4.8	3.04 ÷ 1,000 = 0.00304 $450 \div 10^4 = 0.045$	$98.203 \div 10^5 = 0.00098203$ $493.2 \div 10^6 = 0.0004932$

7. a. 65 = 5 × 13
   b. 75 = 5 × 5 × 3
   c. 82 = 2 × 41

8. 651

9.

a. GCF of 16 and 42 is 2 16 + 42 = 2(8 + 21)	b. GCF of 98 and 35 is 7 98 + 35 = 7 (14 + 5)

10.

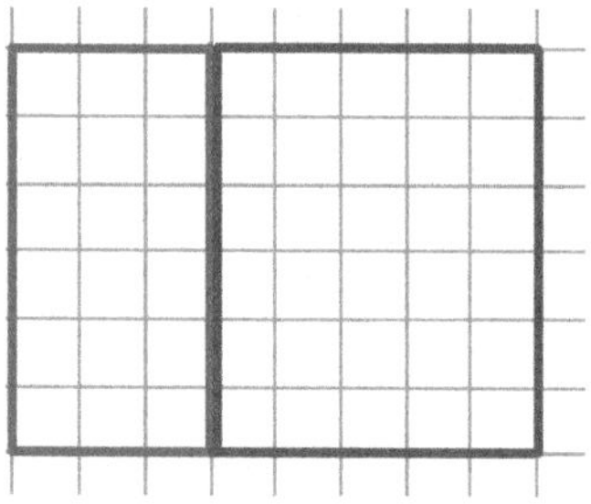

Mixed Review, cont.

11. Calculate the values of y according to the equation $y = 2x - 5$.

x	3	4	5	6	7	8
y	1	3	5	7	9	11

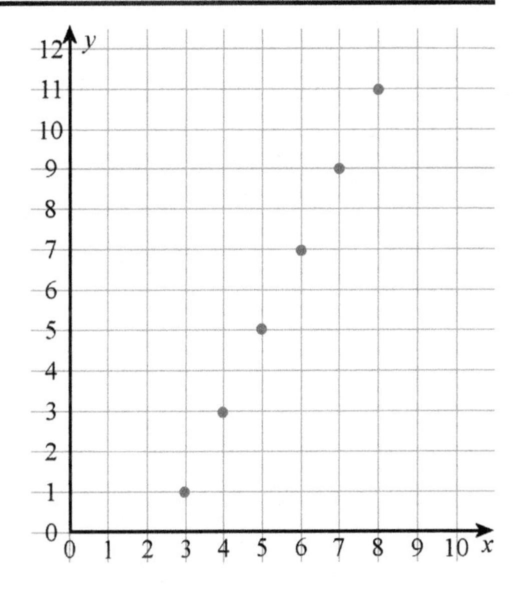

Fractions Review, p. 64

1.

a. $\frac{5}{12} + \frac{4}{12} = \frac{9}{12} = \frac{3}{4}$	b. $\frac{30}{42} + \frac{7}{42} = \frac{37}{42}$	c. $\frac{64}{40} + \frac{35}{40} = \frac{99}{40} = 2\frac{19}{40}$

2.

a. $6\frac{2}{3} \rightarrow 6\frac{4}{6}$ $-\ 2\frac{1}{6} \rightarrow -\ 2\frac{1}{6}$ $4\frac{1}{2}$	b. $7\frac{1}{6} \rightarrow 6\frac{35}{30}$ $-\ 2\frac{3}{5} \rightarrow -\ 2\frac{18}{30}$ $4\frac{17}{30}$	c. $8\frac{9}{11} \rightarrow 8\frac{27}{33}$ $-\ 4\frac{1}{3} \rightarrow -\ 4\frac{11}{33}$ $4\frac{16}{33}$

3.

a. $\frac{3}{4} \times \frac{2}{3} = \frac{6}{12} = \frac{1}{2}$	b. $\frac{1}{5} \times \frac{1}{2} = \frac{1}{10}$	c. $\frac{2}{3} \times \frac{3}{5} = \frac{6}{15} = \frac{2}{5}$

4.

a. 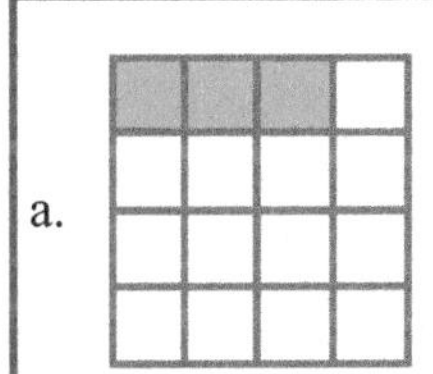 $\frac{1}{4} \times \frac{3}{4} = \frac{3}{16}$

b. 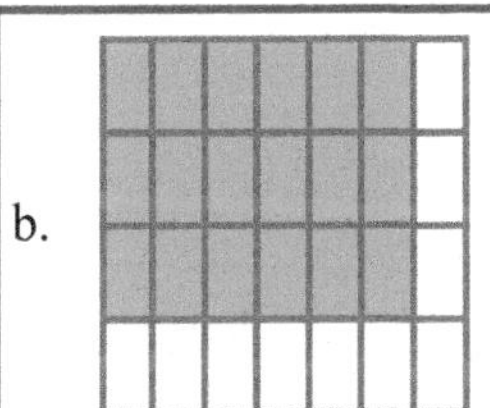 $\frac{3}{4} \times \frac{6}{7} = \frac{9}{14}$

5. a. $\frac{3}{2} \times \frac{1}{5} = \frac{3}{10}$ b. $\frac{1}{5} \times \frac{1}{7} = \frac{1}{35}$ c. $\frac{1}{4} \times \frac{1}{3} = \frac{1}{12}$

6.

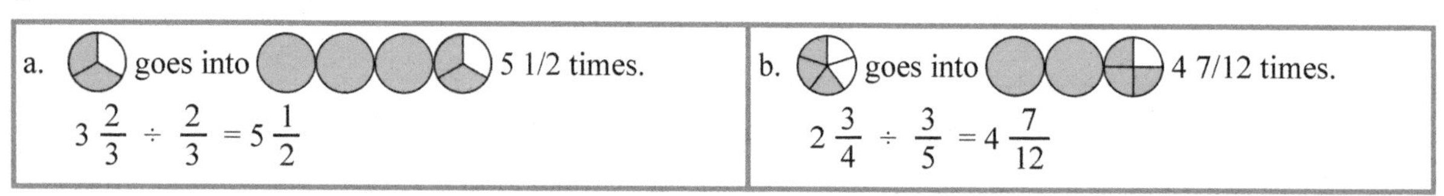

7. Dividing a number by 4 is the same as multiplying it by 1/4.
   Example: 16 ÷ 4 = 4 or 16 × (1/4) = 4.

8. a. 3 1/3 b. 1 3/7 c. 3 3/5

9. a. The area of the garden is 32 13/16 square feet.
   b. The perimeter of the garden is 23 3/4 feet.

10. Answers will vary. Please check the student's answers.
    Example: Three people shared 9/12 of a pie. How much pie did each person get? $\frac{9}{12} \div 3 = \frac{3}{12}$

11. You can cut 28 pieces with a little bit of string left over.
    You can divide to find the answer, but since the two quantities are in different units, one of them needs converted so that both are in feet or that both are in inches. For example, we can change 10 ft into 120 inches, and then divide: 120 in ÷ (4 1/4 in) = 120 in ÷ (17/4 in) = 120 × (4/17) = 480/17 = 28 4/17.

12. One piece was 1 7/8 inches long and the other piece was 13 1/8 inches long.
    The ratio of 1:7 means there are 8 equal parts. Divide 15 inches by 8 to find the length of one part: 15 in ÷ 8 = 15/8 in = 1 7/8 in.

13. The wingspan is 40 feet 3 3/4 inches. Multiply 15 × (32 1/4 in) = 15 × 32 in + 15/4 in = 483 3/4 in = 40 ft 3 3/4 in.

14. There are 36 students in the class. One-sixth of the students is 6 students, so the total number of students is six times that number which is 36.

15. The number is 400. Since two-fifths of the number is 160, one-fifth of it is 80. Multiply that by 5 to get the number.

16. The total kiwi harvest was 82 1/2 pounds.
    The amount given to the son was 11 lb. Two-fifths of the harvest was 33 lb. Therefore, one-fifth of the harvest was 16.5 lb, and the total harvest was five times that, or 82.5 lb.

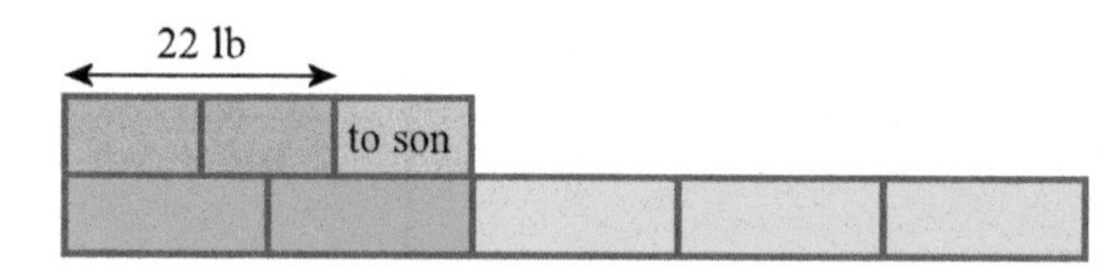

Puzzle corner:
a. 1/240 b. 3/4

Chapter 8: Integers

Integers, p. 71

1. 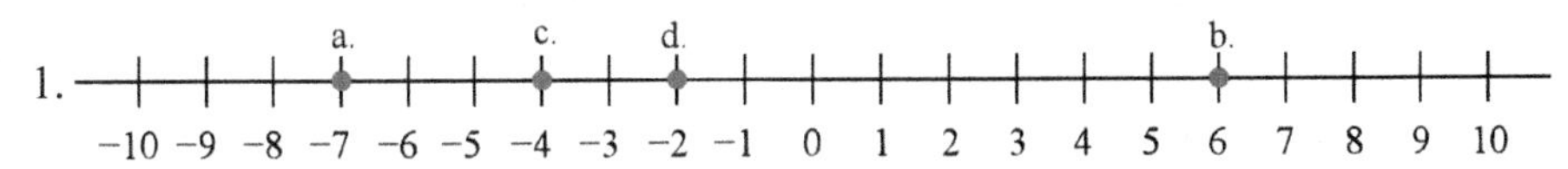

2. a. −\$23 b. +\$250 c. +8,800 m d. −18°C e. −30 ft

3.

before	1°C	2°C	−2°C	−4°C	−12°C	−8°C
change	drops 3°C	drops 7°C	drops 1°C	rises 5°C	rises 4°C	rises 3°C
now	−2°C	−5°C	−3°C	1°C	−8°C	−5°C

4.

a. −2 > −3	b. 8 > −8	c. −3 < 0	d. 4 > −3	e. −5 > −9
f. −10 > −30	g. −4 < 1	h. 0 > −13	i. −2 > −7	j. −11 > −14

5. a. −2 b. 2 c. −2 d. −5

6. a. −5 b. −8 c. −2

7. a. −\$10 < −\$8.
   b. −3 m > −4 m
   c. −10°C < −6°C
   d. \$5 > −\$5
   e. 10°C > −2°C

8.

a. −4 −2 0 4	b. −6 −3 3 5
c. −20 −14 −10 −9	d. −8 −6 −3 0

9. a. 5 b. 12 c. 7 d. 0 e. 68

10. a. −5 = −5 (The opposite of 5 is negative five).
    b. −(−9) = 9 (The opposite of negative nine is nine, or the opposite of the opposite of nine is nine).
    c. −10 = −10 (The opposite of ten is negative ten.)
    d. −0 = 0 (The opposite of zero is zero.)
    e. −(−100) = 100 (The opposite of negative 100 is 100, or the opposite of the opposite of 100 is 100).

11. a. −6
    b. −| 6| = −6
    c. | −6| = 6
    d. |−6| = 6
    e. −(−6) = 6
    f. | −(−6)| = 6

Coordinate Grid, p. 74

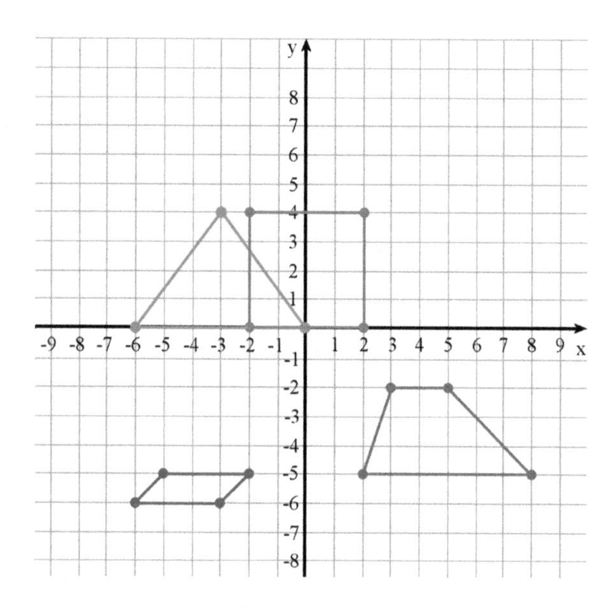

1. A (−6 , 1) B (−3 , 3)
   C (2, 5) D (−4 , −2)
   E (6 , −1) F (−3 , −5)
   G (1 , −5)

2. a. (−2, 4), (2, 4), (−2, 0), (2, 0)
   A square is formed.

   b. (−6, −6), (−5, −5), (−2, −5), (−3, −6)
   A parallelogram is formed.

   c. (2, −5), (3, −2), (5, −2), (8, −5)
   A trapezoid is formed.

   d. (−6, 0), (−3, 4), (0, 0)
   An (isosceles) triangle is formed.

3. The area is 32 square units.

4. The other two vertices are (−5, 4) and (−2, −3).

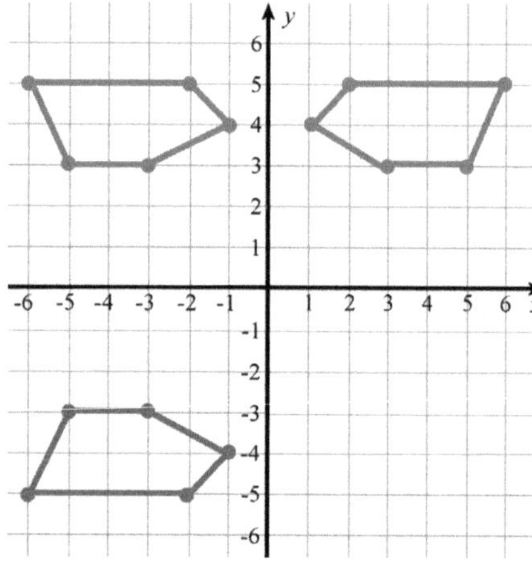

5. a. See the grid on the right.
   b. (6, 5), (5, 3), (3, 3), (1, 4), (2, 5)
   c. The image is mirrored in the *y*-axis.

6. (−6, −5), (−5, −3), (−3, −3), (−1, −4), (−2, −5).
   The image is mirrored in the *x*-axis.

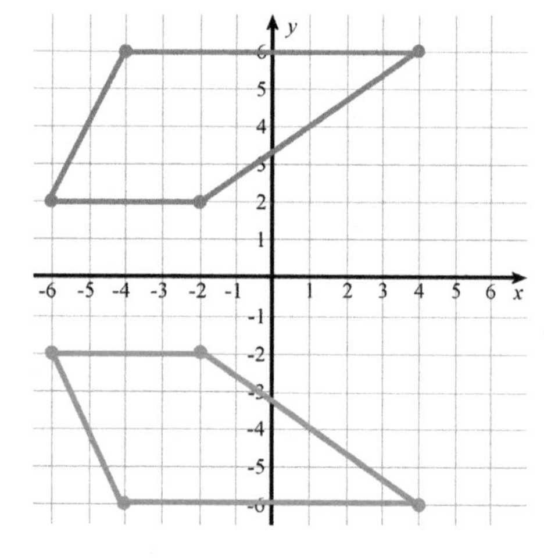

7. a. See the grid on the right.

   b. (−6, −2), (−2, −2), (−4, −6), and (4, −6).
   c. It is a trapezoid.

8. a. (2, −7) b. (−15, −20) c. (−11, 21) d. (34, 19)

9. a. (−3, 9) b. (−22, −20) c. (67, −35) d. (51, 60)

10. a. See the image on the right.

    b. See the image on the right. The coordinates of the reflected figure are (−4, 2), (−7, 2), (−8, 6), and (−5, 6).

    c. See the image on the right. The coordinates of the reflected figure are (4, −2), (7, −2), (8, −6), and (5, −6).

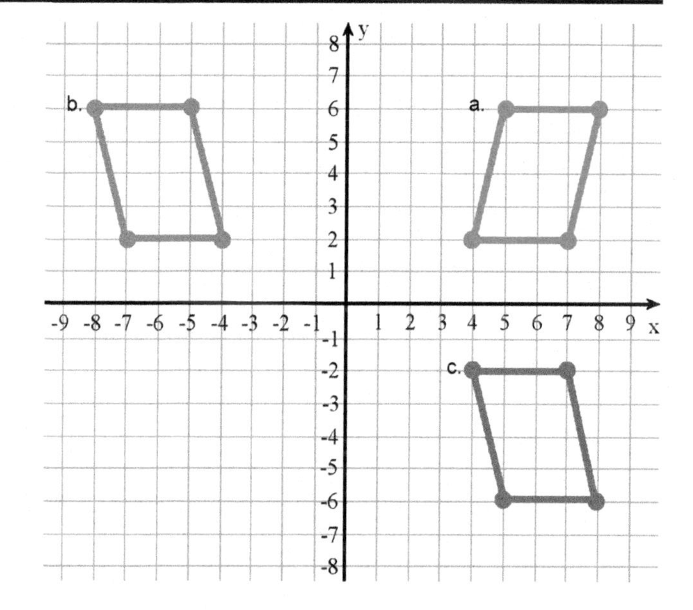

11. The points (−2, 1), (−8, 2), (−6, 3), (−8, 4), and (−2, 5) are the vertices of a pennant.

    a. See the pennant on the right.
    b. See the pennant on the right.
    c. See the pennant on the right.

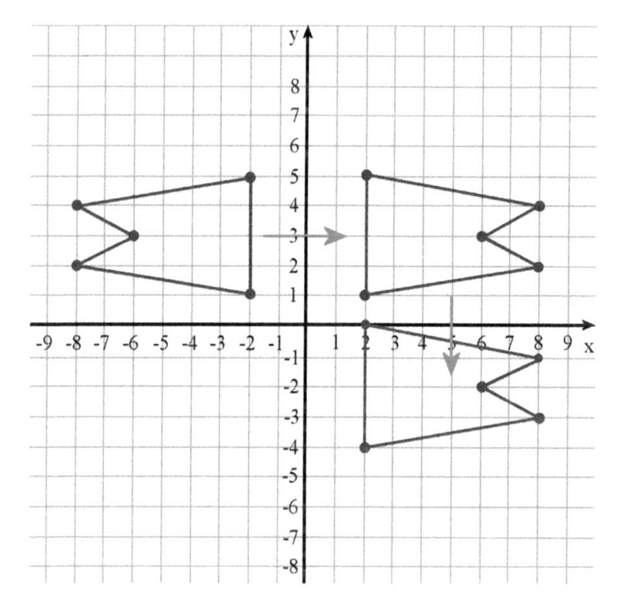

Puzzle corner:
(−7, 5), (−4, 1), and (−2, 4).

Coordinate Grid Practice, p. 78

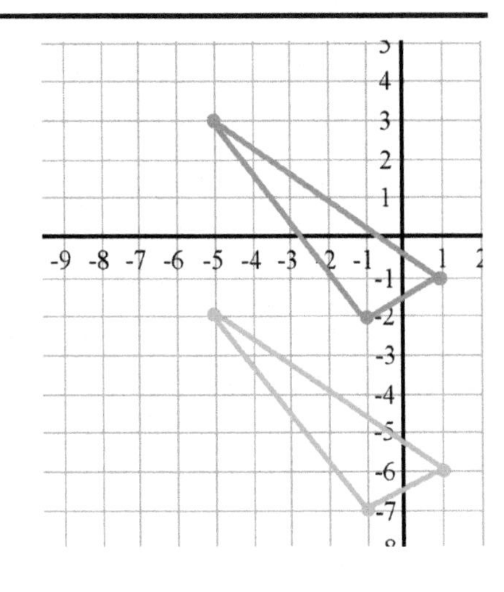

1. a. See the grid on the right.

   b. (−5, −2) → (−5, 3)

   (−1, −7) → (−1, −2)

   (1, −6) → (1, −1)

2.

Point	Direction	New point
(1, 1)	7 units down	(1, −6)
(2, −2)	6 units left	(−4, −2)
(−2, 7)	5 units right	(3, 7)
(−2, −2)	4 units down	(−2, −6)

3. The new coordinates are (3, 2).

4. The coordinates of the original vertices were (−4, 0), (−6, −2), (−4, −6), and (1, −2).

5. a. −3°C and 3°C; the difference is 6°.
   b. 4°C and −6°C; the difference is 10°.
   c. −15°C and −7°C; the difference is 8°.
   d. −4°C and 12°C; the difference is 16°.
   e. −7°C and −28°C; the difference is 21°.
   f. −3°C and 0°C; the difference is 3°.
   g. 6°C and −6°C; the difference is 12°.
   h. 0°C and −13°C; the difference is 13°.

6. Answers will vary. Check the student's answers. For example: "You can add the two numbers without minus signs to find the distance between them. 29 + 28 = 57."

7. You can add the absolute values of the two numbers to find the distance between them. |29| + |28| = 57.

8. Answers will vary. Check the student's answers. For example: "Take the absolute values of the numbers, and subtract the smaller absolute value from the larger. 72 – 49 = 23."

9. a. The distance is 71 units. Add the absolute values of the y-coordinates. 56 + 15 = 71.
   b. The distance is 26 units. Subtract the absolute values of the x-coordinates. 34 − 8 = 26.

10. The perimeter of the rectangle is 270 units.
    The sides of the rectangle are 46 + 22 = 68 units and 50 + 17 = 67 units. The perimeter is 68 + 67 + 68 + 67 = 270 units.

11. a. the school (0, −300)
    b. Grace's house (−250, 250)
    c. the park (−250, −150)
    d. It is 250 + 150 = 400 feet from the park to the gas station.
    e. It is 350 feet from Sophia's house to Grace's house.
    f. The distance from (−200, 100) to (−25, 100) is 175 feet.
    g. The distance from (−300, 250) to (−300, −350) is 600 feet.

12. Please check the student's work.

Puzzle corner:
a. The coordinates of the original vertices were (−6, −2), (−3, 3), (−1, −2), and (2, 3).
   Note: Since Aaron moved his figure UP and to the RIGHT, then originally it was DOWN and to the LEFT from the final points.
b. The figure is a parallelogram.

Addition and Subtraction as Movements, p. 81

1. a. $-8 + 3 = -5$
   b. $-3 + 5 = 2$
   c. $-10 + 4 = -6$
   d. $-4 + 4 = 0$

2. a. $-8 + 2 = -6$ b. $-5 + 4 = -1$

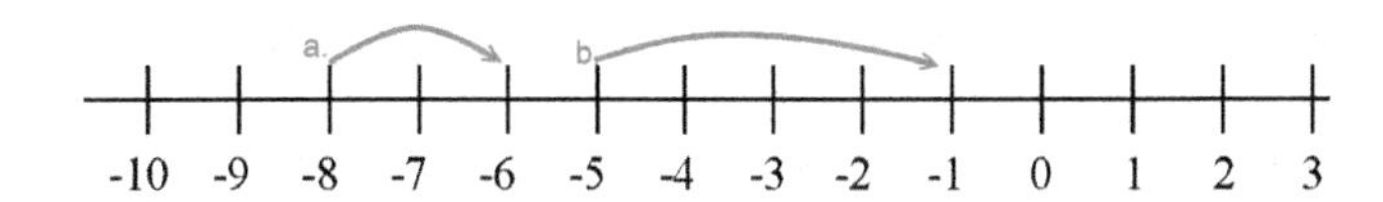

c. $-7 + 5 = -2$ d. $-10 + 12 = 2$

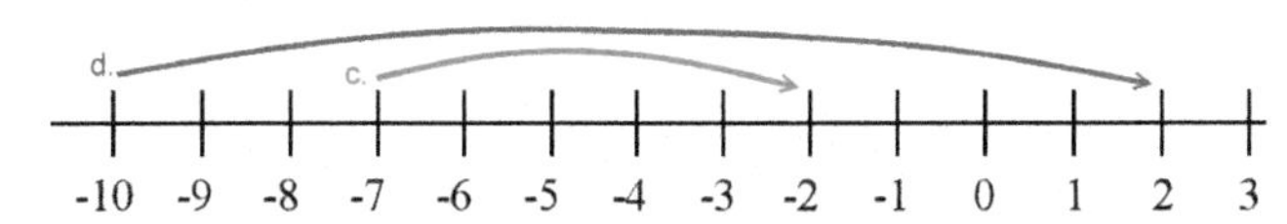

3. a. You are at $^{-}3$. You jump 6 to the right. You end up at 3.
   b. You are at $^{-}8$. You jump 8 to the right. You end up at 0.
   c. You are at $^{-}4$. You jump 7 to the right. You end up at 3.
   d. You are at $^{-}10$. You jump 3 to the right. You end up at $^{-}7$.

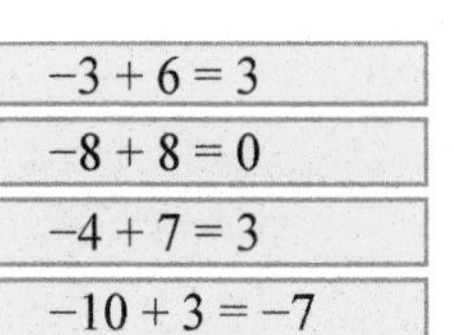
$-3 + 6 = 3$
$-8 + 8 = 0$
$-4 + 7 = 3$
$-10 + 3 = -7$

4. a. $-6 - 3 = -9$
   b. $2 - 4 = -2$
   c. $-3 - 6 = -9$
   d. $0 - 7 = -7$

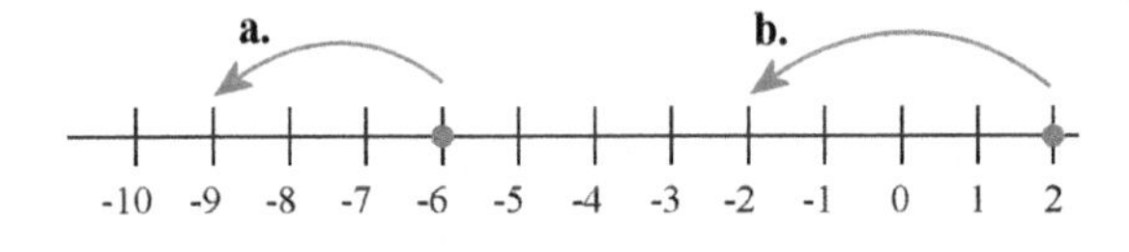

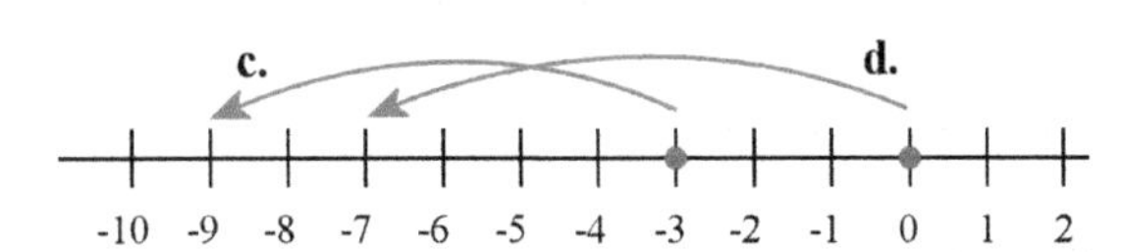

5. a. $1 - 5 = -4$ b. $0 - 8 = -8$

c. $-2 - 4 = -6$ d. $-7 - 3 = -10$

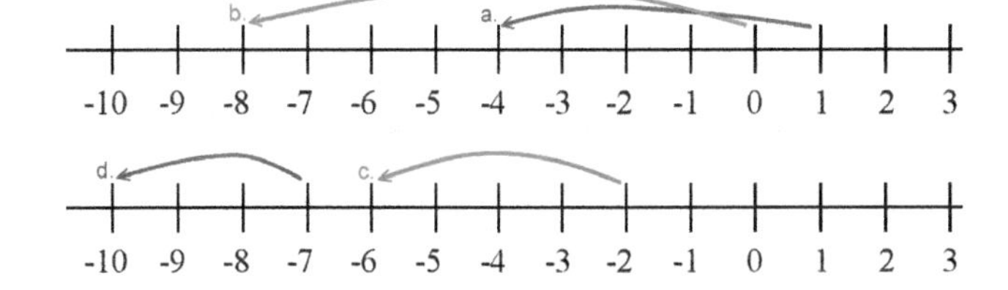

6. a. You are at $^{-}3$. You jump 5 to the left. You end up at -8.
   b. You are at 5. You jump 10 to the left. You end up at -5.
   c. You are at $^{-}5$. You jump 5 to the left. You end up at -10.

$-3 - 5 = -8$
$5 - 10 = -5$
$-5 - 5 = -10$

7.

a.	b.	c.	d.
$3 - 4 = {}^{-}1$	$^{-}2 - 1 = {}^{-}3$	$^{-}4 + 4 = 0$	$^{-}5 + 6 = 1$
$2 - 5 = {}^{-}3$	$^{-}6 - 4 = {}^{-}10$	$^{-}7 + 3 = {}^{-}4$	$^{-}8 + 4 = {}^{-}4$
$5 - 9 = {}^{-}4$	$^{-}7 - 2 = {}^{-}9$	$^{-}12 + 5 = {}^{-}7$	$^{-}6 + 7 = 1$

8.

a. $1 - 5 = {}^{-}4$	c. $^{-}7 + 1 = {}^{-}6$	e. $2 - 7 = {}^{-}5$	g. $^{-}3 + 3 = 0$
b. $3 - 6 = {}^{-}3$	d. $^{-}9 + 8 = {}^{-}1$	f. $0 - 8 = {}^{-}8$	h. $^{-}9 + 18 = 9$

9. He ends up at $^{-}8$.

10. a. $\$5 - \$2 - \$6 = {}^{-}\$3$
    b. James owes $3.

Adding Integers: Counters, p. 84

1. a. −3 b. 2
 c. −9 d. −2
 e. −2 f. −3

2.

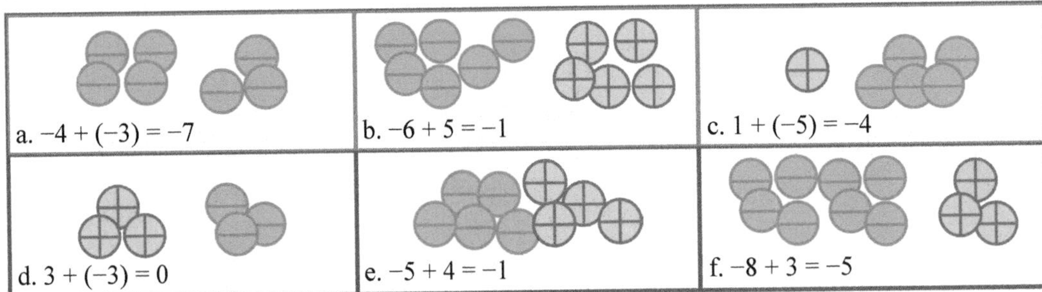

3.

a. $7 + (-8) = {}^{-}1$ $(-7) + 8 = 1$	b. $(-7) + (-8) = {}^{-}15$ $7 + 8 = 15$	c. $5 + (-7) = {}^{-}2$ $7 + (-5) = 2$	d. $50 + (-20) = 30$ $10 + (-40) = {}^{-}30$
e. ${}^{-}2 + {}^{-}4 = {}^{-}6$ ${}^{-}6 + 6 = 0$	f. $10 + {}^{-}1 = 9$ ${}^{-}10 + {}^{-}1 = {}^{-}11$	g. ${}^{-}8 + 2 = {}^{-}6$ ${}^{-}8 + {}^{-}2 = {}^{-}10$	h. ${}^{-}9 + {}^{-}1 = {}^{-}10$ $9 + {}^{-}1 = 8$

4. a. $7 + ({}^{-}5) = 2$
 b. ${}^{-}3 + ({}^{-}11) = {}^{-}14$
 c. $100 + ({}^{-}15) = 85$

5. a. ${}^{-}\$50 + \$60 = \$10$. The balance is now \$10.
 b. ${}^{-}\$20 + ({}^{-}\$15) = {}^{-}\$35$ Hannah's "balance" now is ${}^{-}\$35$.

6. $(-2) + (-6)$, $2 + (-6)$, $(-2) + 6$, $2 + 6$

7.

a. $-3 + (-4) = -7$	b. $-3 + 6 = 3$	c. $3 + (-10) = (-7)$
d. $-7 + (-15) = -22$	e. $2 + (-7) = -5$	f. $5 + (-5) = 0$

8. a. On the number line, $-7 + 4$ is like starting at <u>−7</u>, and moving <u>4</u> steps to the <u>right</u>, ending at −3.
 b. With counters, $-7 + 4$ is like <u>7</u> negatives and <u>4</u> positives added together. We can form <u>4</u> negative-positive pairs that cancel, and what is left is <u>3</u> negatives.

9.

a. $4 + (-10) = -6$ $-6 + 8 = 2$	b. $-8 + (-8) = -16$ $7 + (-8) = -1$	c. $-5 + (-7) = -12$ $12 + (-5) = 7$	d. $11 + (-2) = 9$ $-10 + 20 = 10$

10. a. Answers will vary. Please check the student's answers.
 Examples: $4 + (-4) = 0$, $8 + (-4) = 4$, $-8 + (-4) = -12$, $-12 + (-4) = -16$.
 b. If $x = 4$, then $x + (-4) = 0$.

11.

a. $3 - 2 = 1$ $3 - 3 = 0$ $3 - 4 = {}^{-}1$ $3 - 5 = {}^{-}2$ $3 - 6 = {}^{-}3$	b. ${}^{-}7 - 0 = {}^{-}7$ ${}^{-}7 - 1 = {}^{-}8$ ${}^{-}7 - 2 = {}^{-}9$ ${}^{-}7 - 3 = {}^{-}10$ ${}^{-}7 - 4 = {}^{-}11$	c. ${}^{-}5 + 0 = {}^{-}5$ ${}^{-}5 + 1 = {}^{-}4$ ${}^{-}5 + 2 = {}^{-}3$ ${}^{-}5 + 3 = {}^{-}2$ ${}^{-}5 + 4 = {}^{-}1$	d. ${}^{-}6 + 6 = 0$ ${}^{-}6 + 7 = 1$ ${}^{-}6 + 8 = 2$ ${}^{-}6 + 9 = 3$ ${}^{-}6 + 10 = 4$

Subtracting a Negative Integer, p. 87

1. a. −7 − (−3) = −4
   b. −7 − (−6) = −1

2. a. −2 − 6 = −8
   b. −2 − (−6) = 4

3.

a. The distance between 3 and −7 is 10. Subtraction: 3 − (−7) = 10	b. The distance between −3 and −9 is 6. Subtraction: −3 − (−9) = 6
c. The distance between −2 and 10 is 12. Subtraction: 10 − (−2) = 12	d. The distance between −11 and −20 is 9. Subtraction: −11 − (−20) = 9

4.

a. −8 − (−4) = −4 8 − (−4) = 12 −8 + (−4) = −12 8 + (−4) = 4	b. −1 − (−5) = 4 1 − (−5) = 6 −1 − 5 = −6 1 − 5 = −4	c. 12 − (−15) = 27 −12 + 15 = 3 −12 − 15 = −27 12 + (−15) = −3

5.

a.		b.	
10 − (−3)	10 − 3	−9 + 2	−9 + (−2)
10 + (−3)	10 + 3	−9 − 2	−9 − (−2)

6. a. 90 ft − 105 ft = −15 ft
   b. \$25 − \$40 = −\$15. Matt will owe his friend \$15.

Puzzle corner: −1 + (−2) − (−3) − 4 = − 4

Add and Subtract Roundup, p. 89

1.

a. −4 + 6 = 2 4 + (−6) = −2	b. −3 + 7 = 4 −3 + (−7) = −10	c. 8 + (−9) = −1 (−8) + (−9) = −17

2.

a. (−4) − 6 = − 10 4 − (−6) = 10	b. −3 − 7 = − 10 −3 − (−7) = 4	c. 8 − (−9) = 17 (−8) − (−9) = −1

3.

a. −4 + (−6) = −10 4 + (−6) = −2	b. 6 + (−6) = 0 −6 − (−2) = −4	c. 5 − 7 = −2 4 + (−3) = 1

4. a. You are at $^-6$. You jump 5 to the right. You end up at $^-1$. — $^-6 + 5 = {}^-1$
   b. You are at $^-2$. You jump 7 to the right. You end up at 5. — $^-2 + 7 = 5$
   c. You are at 4. You jump 3 to the left. You end up at 1. — $4 - 3 = 1$
   d. You are at 0. You jump 12 to the left. You end up at $^-12$. — $0 - 12 = {}^-12$
   e. You are at 7. You jump 22 to the left. You end up at $^-15$. — $7 - 22 = {}^-15$
   f. You are at $^-7$. You jump 22 to the right. You end up at 15. — $^-7 + 22 = 15$

Add and Subtract Roundup, cont.

5.

Situation	Addition or Subtraction Sentence
... Now it is at an elevation of ⁻100 m.	⁻65 m − 35 m = ⁻100 m
... Now Henry owes ⁻$210 on his credit card.	⁻$150 + $60 + (⁻$120) = ⁻$210
... Her money situation is now ⁻$140.	⁻$180 + $40 = ⁻$140
... Now the temperature is ⁻1°C.	2°C − 5°C + 2°C = ⁻1°C

6.

L $\underline{-4} + (-8) = -12$
O $3 + (-11) = \underline{-8}$
C $-2 - \underline{3} = -5$
S $-2 - (-4) = \underline{2}$
S $-2 - \underline{(-7)} = 5$

R $-24 - (-25) = \underline{1}$
A $\underline{-2} + 4 = 2$
I $\underline{7} + (-7) = 0$
O $-3 + 5 + (-5) = \underline{-3}$
A $7 - \underline{8} = -1$

M $(-144) + 150 = \underline{6}$
H $77 - 90 = \underline{-13}$
I $-4 + (-15)\ \underline{-19}$
C $-7 + \underline{9} = 2$

What game do cows play at parties?

6	−8	−3	2	−19	9	8	−4
M	O	O	S	I	C	A	L

2	−13	−2	7	1	−7
C	H	A	I	R	S

Graphing, p. 91

1. a. $y = x + 4$

x	−9	−8	−7	−6	−5	−4	−3	−2
y	−5	−4	−3	−2	−1	0	1	2

x	−1	0	1	2	3	4	5	6
y	3	4	5	6	7	8	9	10

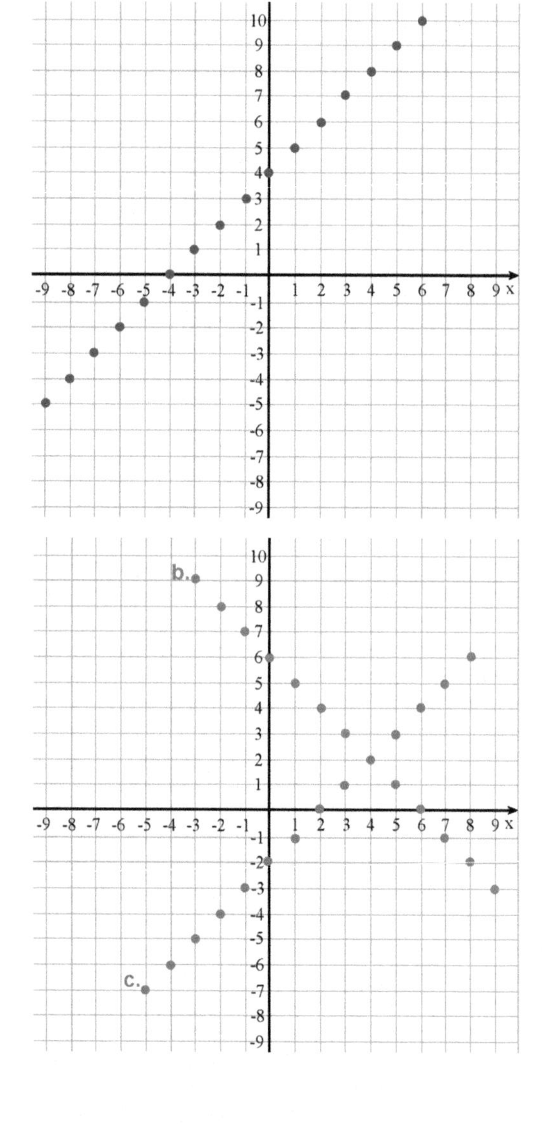

b. $y = 6 - x$

x	−3	−2	−1	0	1	2	3
y	9	8	7	6	5	4	3

x	4	5	6	7	8	9
y	2	1	0	−1	−2	−3

c. $y = x - 2$

x	−5	−4	−3	−2	−1	0	1	2
y	−7	−6	−5	−4	−3	−2	−1	0

x	3	4	5	6	7	8	9
y	1	2	3	4	5	6	7

2. a.

add 1	x	−5	−4	−3	−2	−1
add 2	y	−8	−6	−4	−2	0

add 1	x	0	1	2	3	4
add 2	y	2	4	6	8	10

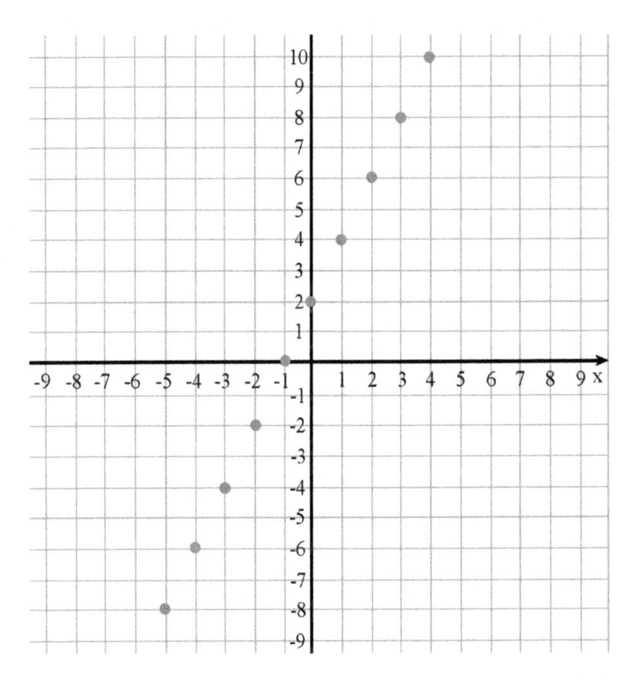

b.

add 2	x	−8	−6	−4	−2	0
subtract 3	y	9	6	3	0	−3

add 2	x	2	4	6	8	10
subtract 3	y	−6	−9	−12	−15	−18

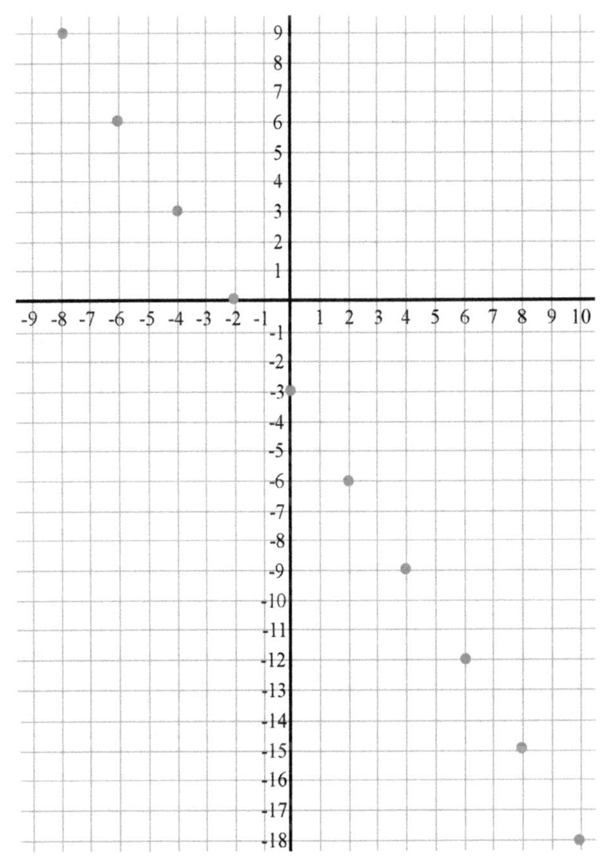

c. Check the student's work. Answers will vary.

3. a.

t	0	1	2	3	4	5	6	7	8	9	10	11	12	13
a	−200	−185	−170	−155	−140	−125	−110	−95	−80	−65	−50	−35	−20	−5

b.

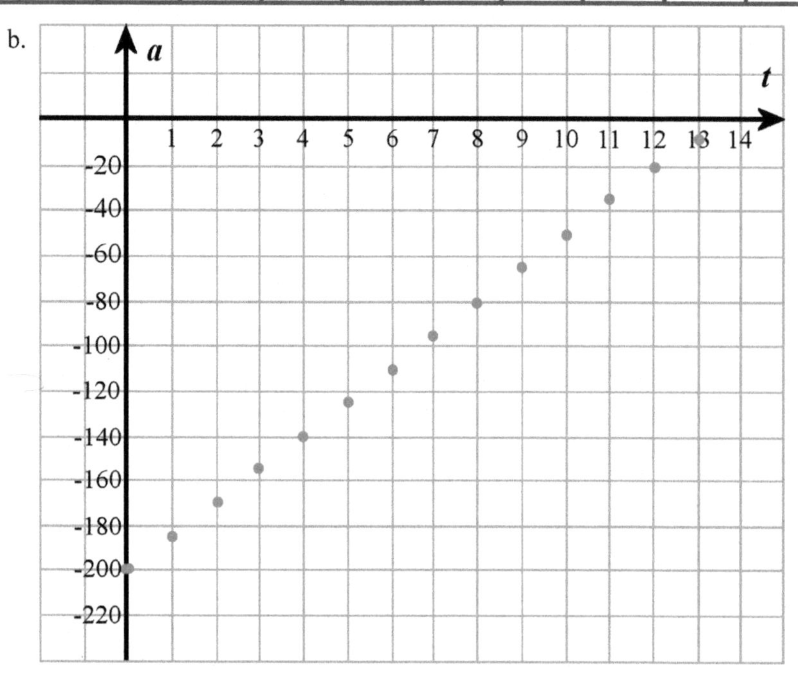

c. Natalie will finish paying off her debt in thirteen weeks.
d. The graph shows how much she has left to pay each week.
It shows she only has to pay $5 the thirteenth week.

4. a.

x	−2	−1	0	1	2	3	4	5
y	−7	−5	−3	−1	1	3	5	7

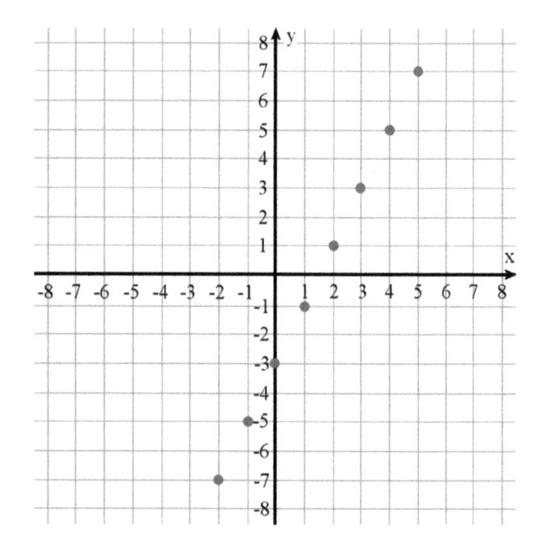

b.

x	−8	−7	−6	−5	−4	−3	−2
y	−19	−17	−15	−13	−11	−9	−7

Graphing, cont.

5. a.

months	*t*	0	1	2	3	4	5	6	7	8	9	10	11	12
balance	*a*	\$500	\$465	\$430	\$395	\$360	\$325	\$290	\$255	\$220	\$185	\$150	\$115	\$80

b.

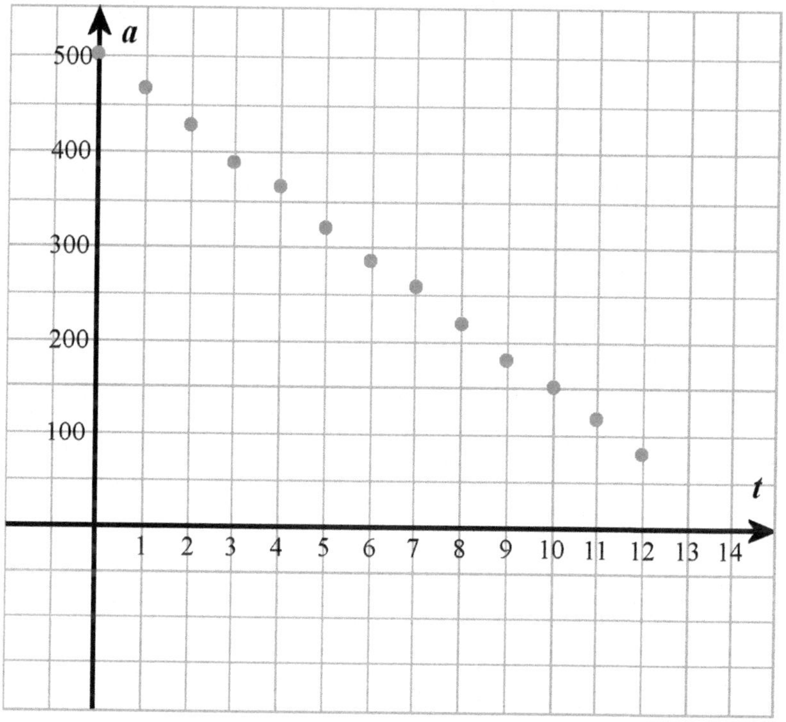

c. After the fourteenth monthly payment, there will not be enough money left to make the next payment. There will only be \$10 left. So at the time of the fifteenth monthly payment, her account will go negative.

d. The balance after the fourteenth monthly payment will be \$10.

Puzzle corner:
When x is 100, y will be -194. ($y = 6 - 2x$)

Mixed Review, p. 95

1. a. The unknown is Shelly's age so choose a variable for it. Let s be Shelly's age. We get the equation $54 - 12 = s$. Solution: $s = 42$

   b. The unknown is the number of tulips so choose a variable for it, such as n. The equation is $n \times \$2.15 = \$45.15$ or you can write this as $2.15n = 45.15$. Solution: $n = \$45.15 \div 2.15 = 21$. Bob bought 21 tulips for his wife.

2. $512 = 2^9$

3.

a. $\frac{5}{5} \div \frac{4}{5} = 1\frac{1}{4}$	b. $\frac{8}{8} \div \frac{3}{8} = 2\frac{2}{3}$
c. $\frac{3}{4} \times \frac{2}{3} = \frac{1}{2}$	d. $\frac{2}{9} \times \frac{3}{4} = \frac{1}{6}$

4. a. Fifty packages of dominoes weigh 9 lb 6 oz. Multiply to find the total weight: 50×3 oz = 150 oz = 9 lb 6 oz.

   b. The quality dominoes weigh 14 lb 10 oz more than the box with cheap dominoes.

Mixed Review, cont.

5.

a. $\frac{23}{8} \div \frac{2}{5} = 7\frac{3}{16}$	b. $\frac{4}{1} \div \frac{11}{6} = 2\frac{2}{11}$
c. $\frac{5}{1} \div \frac{2}{7} = 17\frac{1}{2}$	d. $\frac{101}{10} \div \frac{3}{4} = 13\frac{7}{15}$

6. Annabelle can type 35 words in a minute, so in 15 minutes she can type 15 × 35 = 525 words.

7. Mom is 42 years old. From the model on the right, we can see that each block is 6 years. Therefore, Mom is 7 × 6 years = 42 years old.

Mom

Dad 6

8. The area is 360 cm^2.
The aspect ratio of 5:2 means the two sides make up 7 "parts" and the total perimeter makes 14 parts. Divide the perimeter 84 cm by 14 to find the length of one part:
84 cm ÷ 14 = 6 cm. The sides are 5 × 6 cm = 30 cm and 2 × 6 cm = 12 cm. The area is 30 cm × 12 cm = 360 cm^2.

9. Keith paid approximately 22% of his salary in taxes. His total salary was $414 + $1459 = $1873. The percentage is $414 / $1873 = 0.22103577...

10. a. 180 km : 1 hr
b. 1 kg for $0.70
c. 24 miles per 1 gallon

11.

a. $\frac{\overset{1}{\cancel{5}}}{\underset{3}{\cancel{36}}} \times \frac{\overset{2}{\cancel{24}}}{\underset{9}{\cancel{45}}} = \frac{2}{27}$	b. $\frac{\overset{\overset{1}{\cancel{2}}}{\cancel{16}}}{\underset{\underset{3}{\cancel{6}}}{\cancel{30}}} \times \frac{\overset{5}{\cancel{25}}}{\underset{3}{\cancel{24}}} = \frac{5}{9}$	c. $\frac{\overset{\overset{1}{\cancel{2}}}{\cancel{14}}}{\underset{5}{\cancel{25}}} \times \frac{\overset{7}{\cancel{35}}}{\underset{\underset{3}{\cancel{6}}}{\cancel{42}}} = \frac{7}{15}$

Integers Review, p. 97

1. a. −1 > −7 b. 2 > −2 c. −6 < 0 d. 8 > −3 e. −8 < −3

2. a. −6 −2 0 2	b. −14 −11 −8 −7

3. a. −$12 > −$18
b. −5°C < 2°C
c. 16 m > −6 m

4. a. 11 b. 2 c. 0 d. 19 e. −7

5. a. −9 + 6 = −3 b. −2 + 5 = 3

c. −3 − 5 = −8 d. 2 − 8 = −6

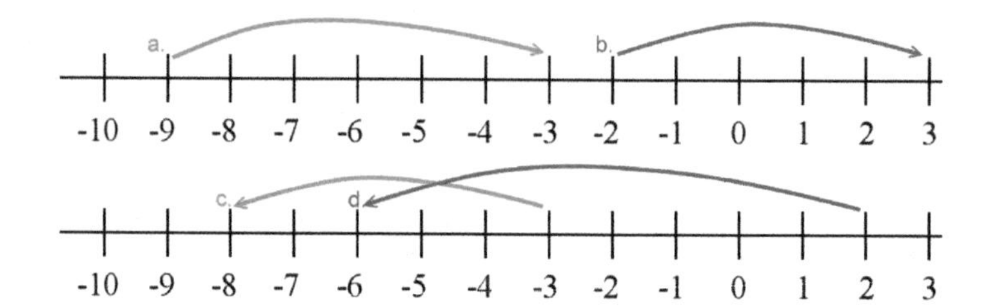

6. a. You are at $^{-}10$. You jump 6 to the right. You end up at $^{-}4$. $\quad$ $^{-}10 + 6 = {}^{-}4$

   b. You are at $^{-}5$. You jump 8 to the right. You end up at 3. $\quad$ $^{-}5 + 8 = 3$

   c. You are at 3. You jump 7 to the left. You end up at $^{-}4$. $\quad$ $3 - 7 = {}^{-}4$

   d. You are at $^{-}11$. You jump 3 to the left. You end up at $^{-}14$. $\quad$ $^{-}11 - 3 = {}^{-}14$

7.

a. $2 + (-8) = {}^{-}6$ $(-2) + 8 = 6$	b. $-2 + (-9) = {}^{-}11$ $2 - 8 = {}^{-}6$	c. $1 + (-7) = {}^{-}6$ $-4 - 5 = {}^{-}9$	d. $5 - (-2) = 7$ $-3 - (-4) = 1$

8. a. May has \$35. She wants to purchase a guitar for \$85.
   That would make her money situation be $^{-}$\$50.

   b. A fish was swimming at the depth of 6 ft. Then he sank 2 ft.
   Then he sank 4 ft more. Now he is at the depth of $^{-}12$ ft.

   c. Elijah owed his dad \$20. Then he borrowed another \$10.
   Now his balance is $^{-}$\$30.

   d. The temperature was −13°C and then it rose 5°.
   Now the temperature is −8°C.

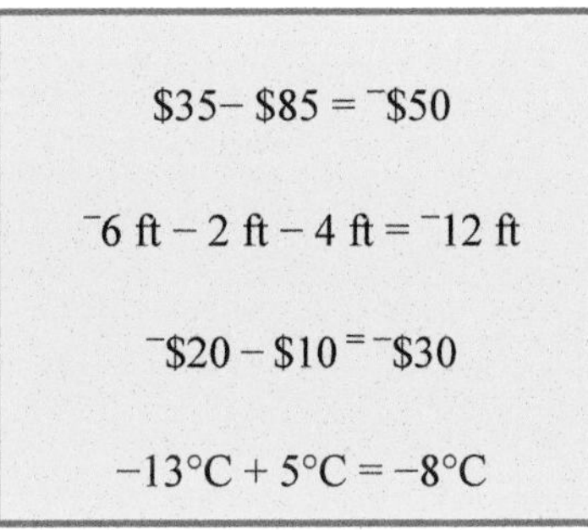

9. a. $|-17|$
   b. $-(-11)$

10. d. balance < −\$50

11.

x	−5	−4	−3	−2	−1	0	1	2
y	9	8	7	6	5	4	3	2

x	3	4	5	6	7	8	9
y	1	0	−1	−2	−3	−4	−5

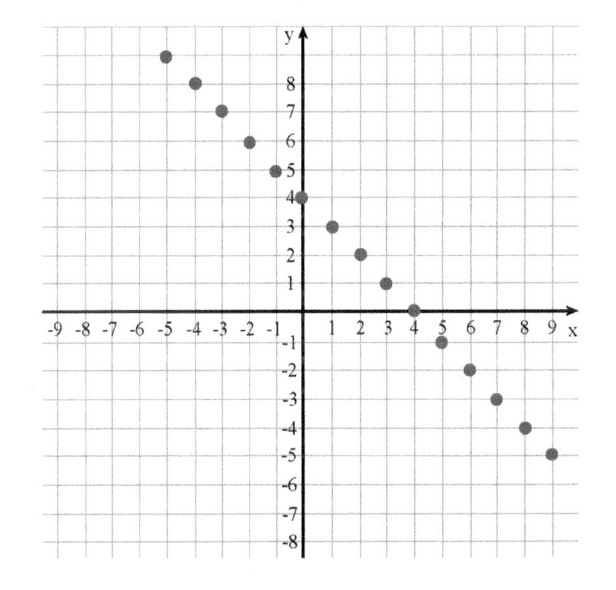

Integers Review, cont.

12.

a. $-2 + (-6) = -8$ $3 + (-5) = -2$	b. $4 + (-4) = 0$ $-6 - 6 = -12$	c. $5 - 7 = -2$ $3 + (-2) = 1$

13. (−9, −6), (−6, −6), (−9, −3), and (−3, 0).

14. a. The distance is 12 + 15 = 27 units.
    b. The distance is 21 − 15 = 6 units.

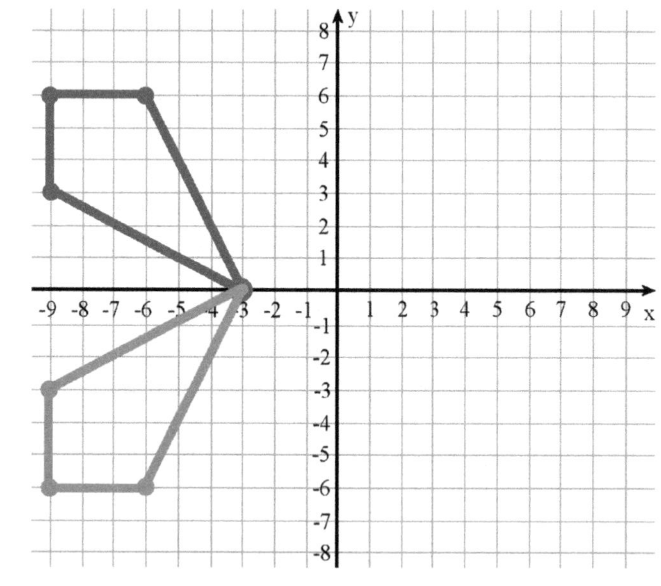

15. a. The points (−7, −3), (−1, −7), (−1, −1), and (−4, −6)

    b. (−7, −3) → (7, −3)

    (−1, −7) → (1, −7)

    (−1, −1) → (1, −1)

    (−4, −6) → (4, −6)

    c. (7, 4) (1, 0) (1, 6) (4, 1)

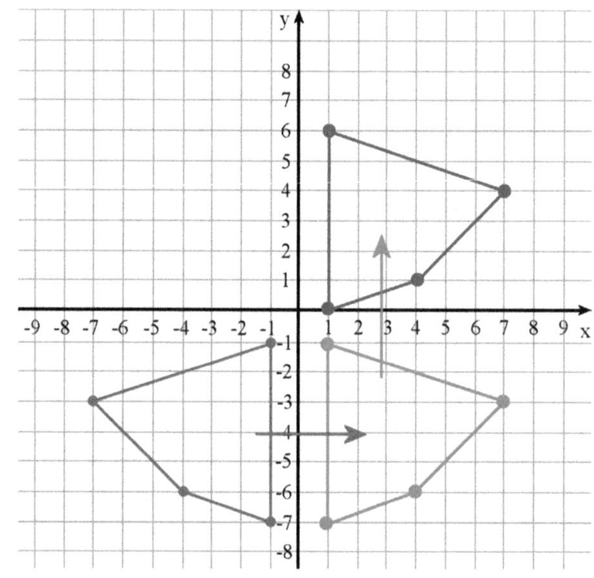

Chapter 9: Geometry

Quadrilaterals Review, p. 104

Teaching box:

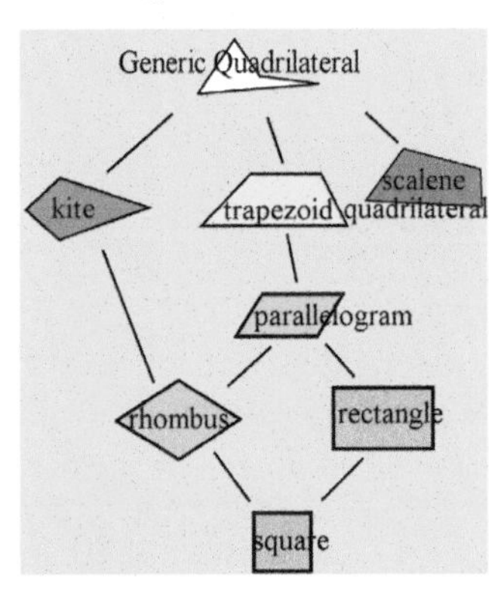

1. a. rhombus
   b. parallelogram
   c. trapezoid
   d. kite
   e. square

2. a. yes
   b. yes
   c. no

3. a. and b. are below.
   Answers to part c will vary; check the student's answers.

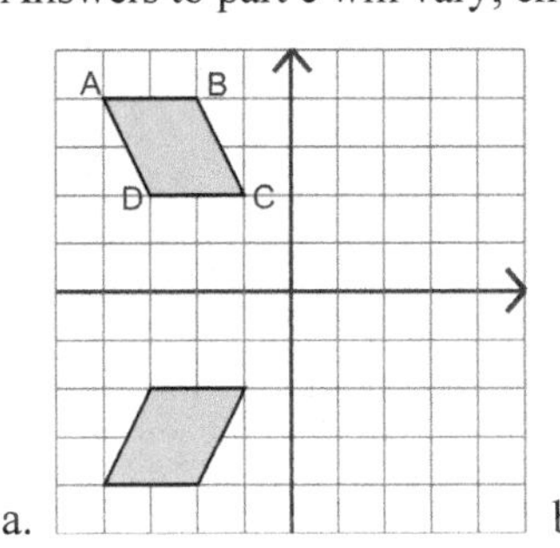

a.

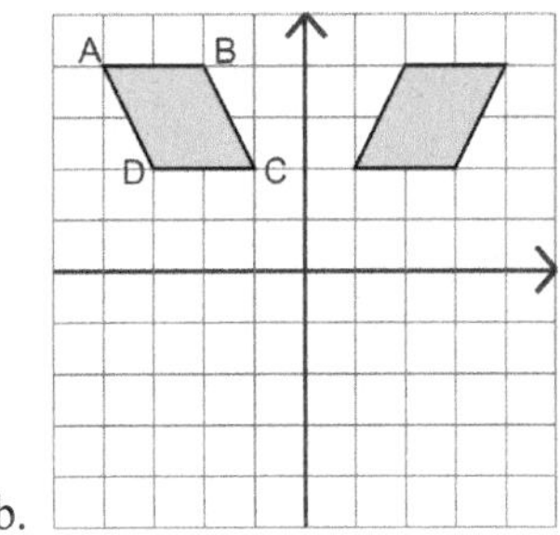

b.

4. There are many such trapezoids. For example (image not to scale):

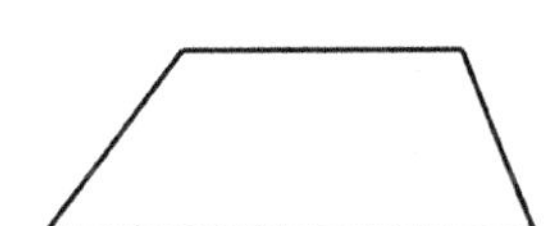

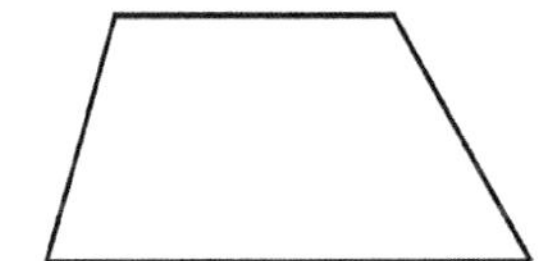

5. Check the student's work by measuring the two angles and the lengths of the sides. The angles should be 30° and 150° and the sides 6 cm. The image here is not to scale.

6. There are four such trapezoids, two basic ones and their mirror images:

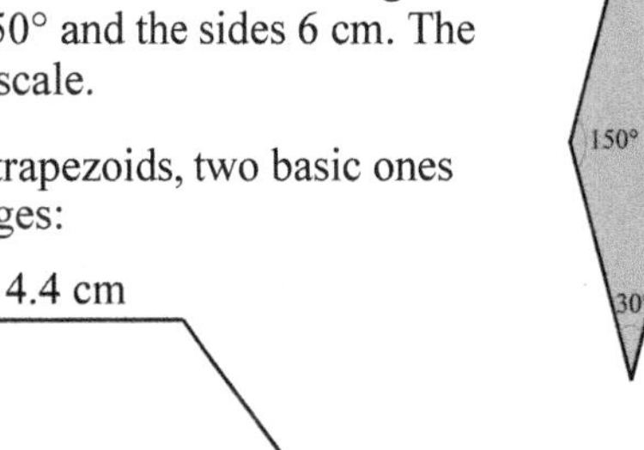

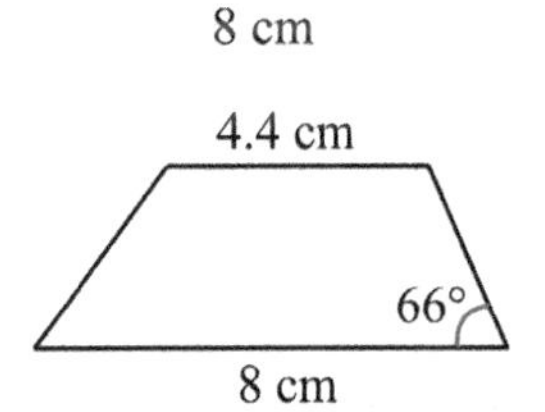

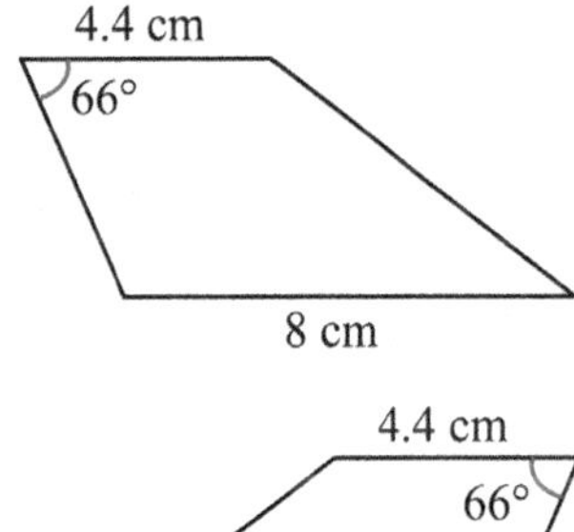

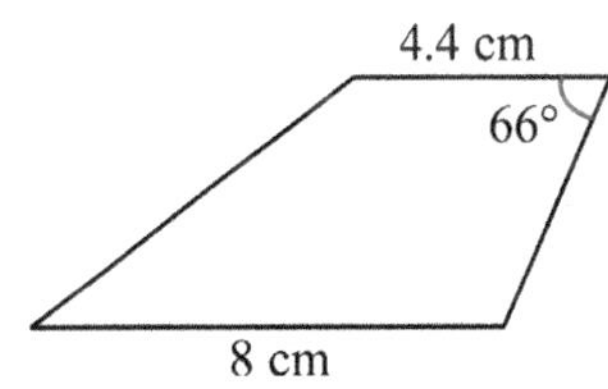

Drawing Problems, p. 106

1.

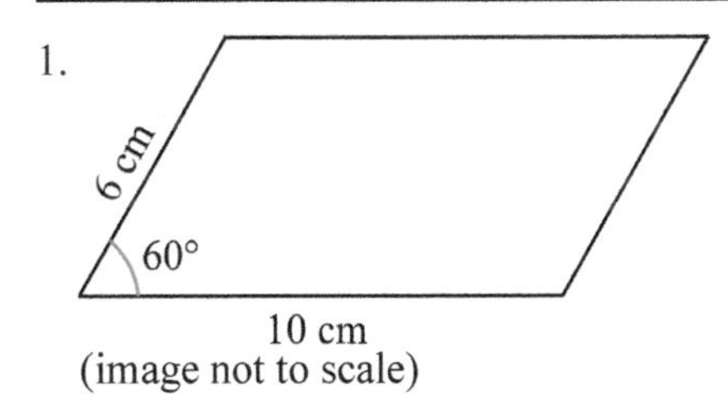

(image not to scale)

2. Start out by drawing one of the 7.6-cm sides. Then draw the 78° angle, and then the other 7.6-cm side. The triangle does not have to be oriented as the image below, with the top angle appearing "at the top".

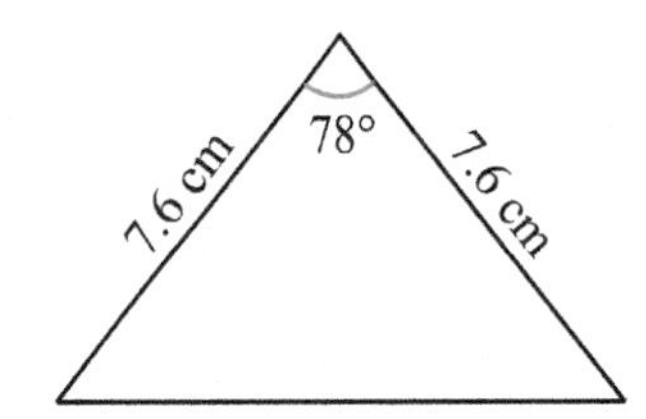

Drawing Problems, cont.

3.

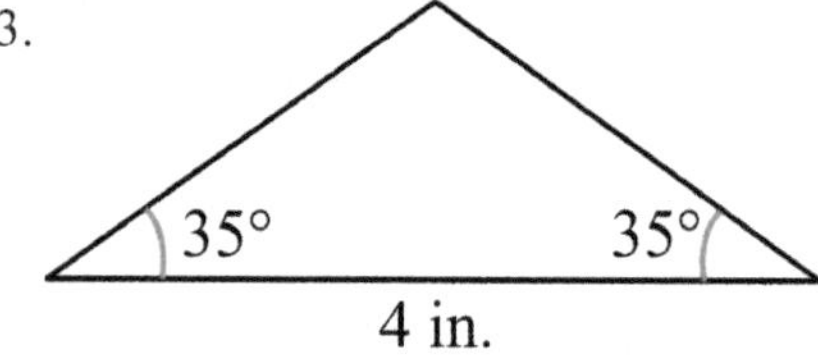

The top angle is 110°. The other two sides are 2.44 inches or 2 7/16 inches.

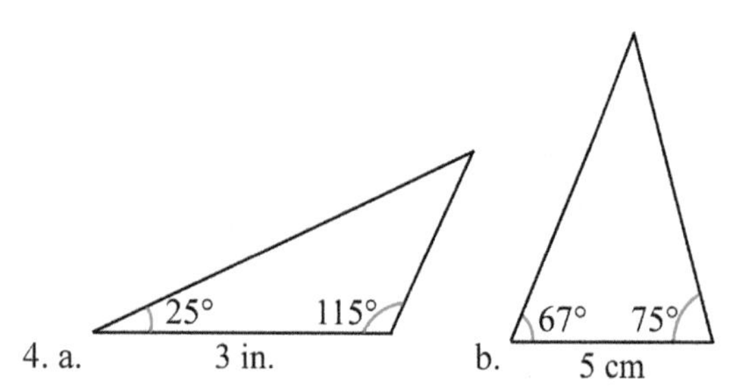

4. a. b.

5. Measure any two angles and a side between those angles from the triangle. First, draw the side you measured. Then draw the two angles, and continue the sides of the angles until they meet.

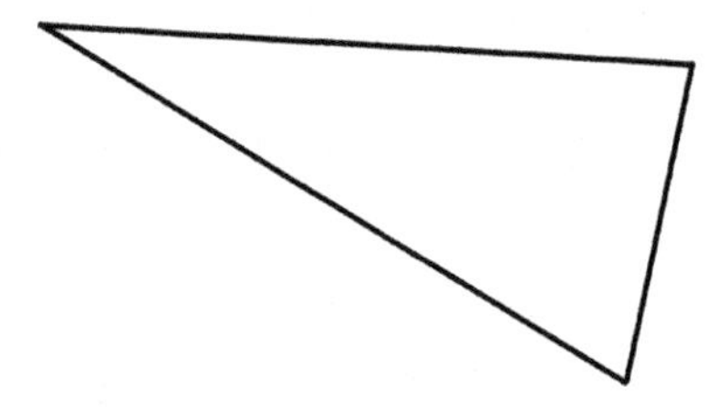

6. There are several different trapezoids you can draw, for example:

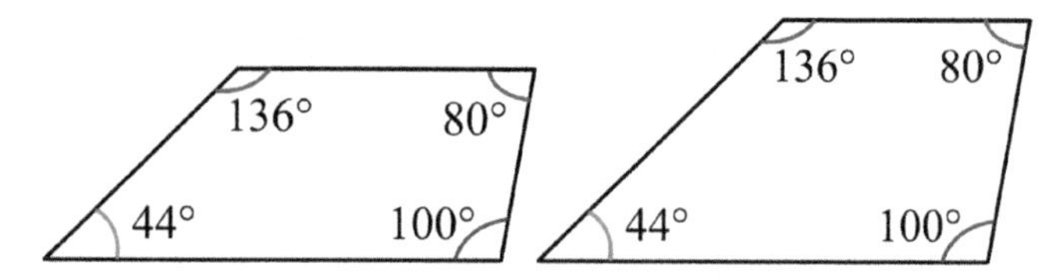

Area of Right Triangles, p. 108

1. a. 8 square units b. 6 square units
   c. 5 square units d. 6 square units
   e. 8 square units f. 2 square units
   g. 2 1/2 square units h. 5 square units

To find the area of a right triangle, **multiply the lengths of the two sides** that are *perpendicular* to each other (in other words, the two that form the right angle). Then take **half of that**.

This works because the area of a right triangle is exactly half of the area of a certain rectangle.

2. a. The area is 3 cm^2.

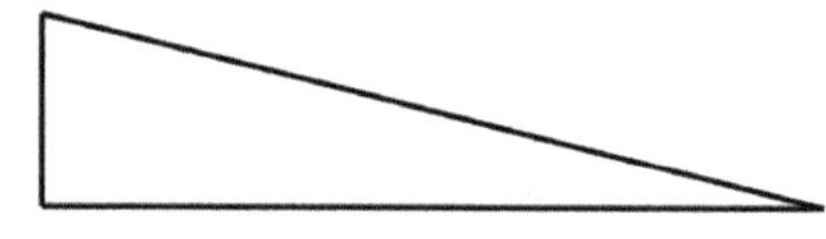

b. The area is (2 1/2) × (1 1/4) ÷ 2 = (5/2) × (5/4) ÷ 2 = 25/8 ÷ 2 = 25/16 = 1 9/16 sq. in.

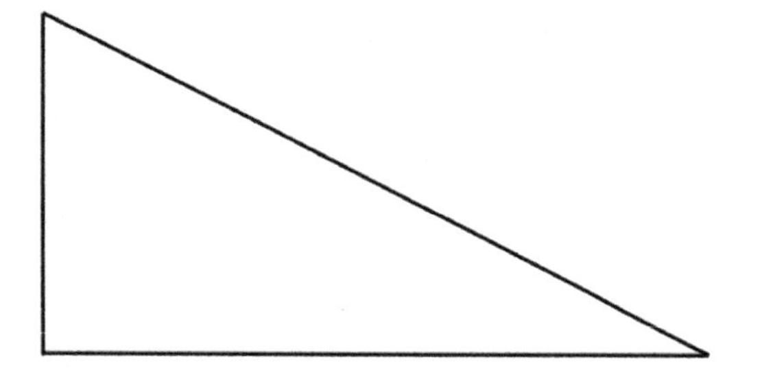

3. a. 3 + 2 + 3 = 8 square units
   b. 4 + 4 + 4 + 4 + 4 = 20 square units
   c. 4 + 1 + 1 + 1 + 1 + 1 + 1 + 1 + 1 = 12 square units

4. You can draw several. The product of the two sides just needs to be 26. So the lengths of the sides could be, for example, 2 cm and 13 cm, or 4 cm and 6.5 cm, or 5 cm and 5.2 cm.

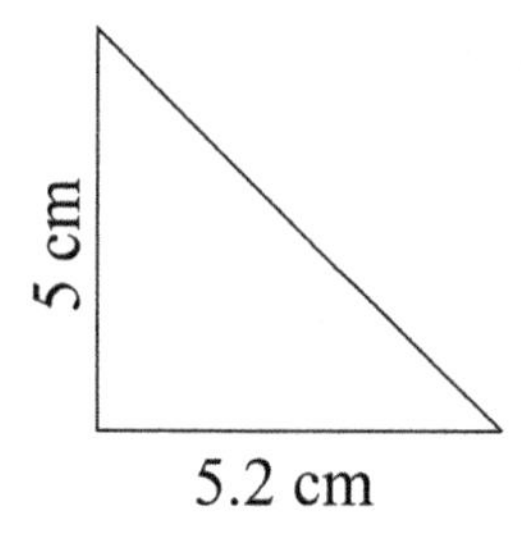

5. Answers will vary. Check the student's work. For example:

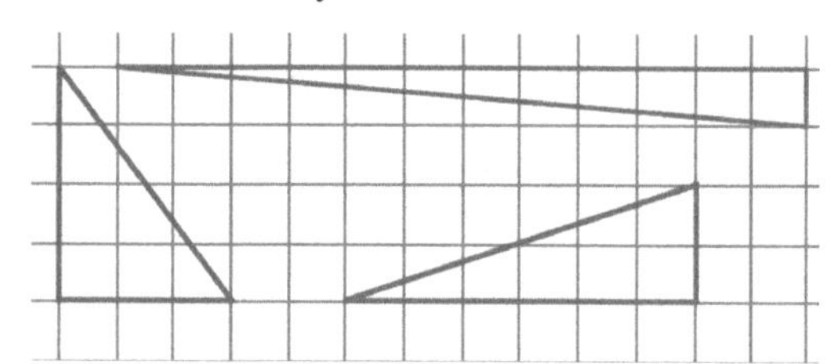

Area of Parallelograms, p. 110

1. 3 units × 6 units = 18 square units

2. a. 2 units × 5 units
   = 10 square units
   b. 3 units × 6 units
   = 18 square units
   c. 4 units × 4 units
   = 16 square units
   d. 5 units × 1 unit
   = 5 square units

3. a. 25 sq. units
   b. 3 sq. units
   c. 6 sq. units
   d. 9 sq. units
   e. 25 sq. units

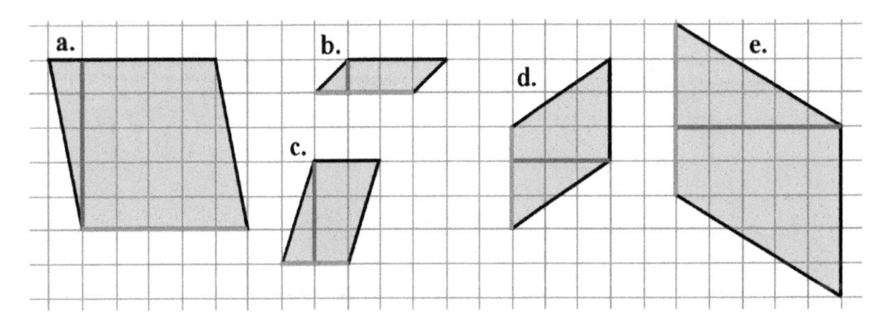

4. a.

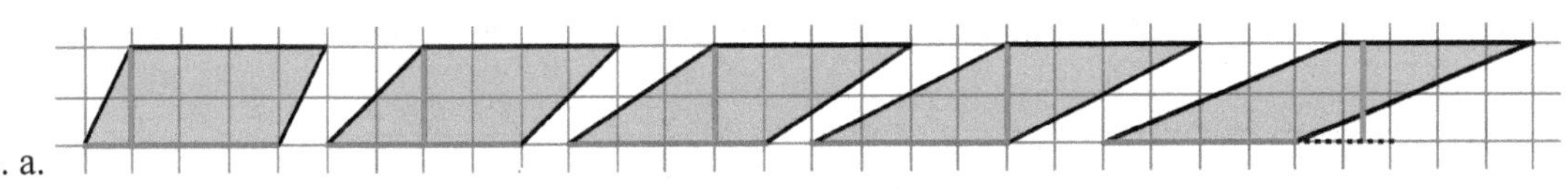

b. They all have the same area -- 8 square units. That is because they all have the same altitude (2 units) and the same base (4 units).

5. It is not possible to draw all possible parallelograms with the area of 12 square units on the grid. Basically, first choose the length of the base and the altitude to make an area of 12 square units—for example, 6 × 2, 3 × 4, 1 × 12, 3.6 × 3 1/3, and so on. Then, for each base and altitude, there is an infinite variety of parallelograms, each with a different amount of "slant." Here are some examples:

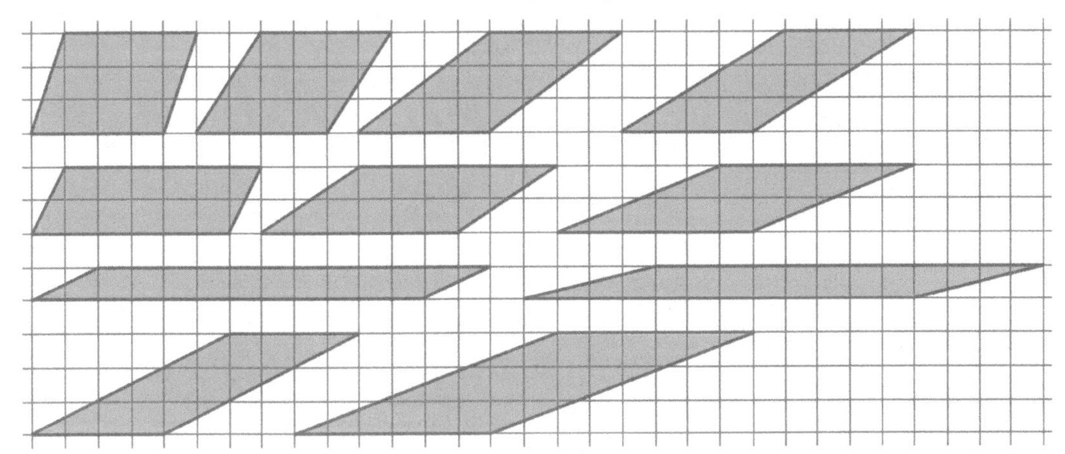

6. Note that the altitude does not have to be drawn from a vertex; it just needs to be perpendicular to the base and to extend from the base to the opposite side.
   a. Base 6.0 cm × altitude 2.9 cm = area 17.4 $\text{cm}^2 \approx 17\ \text{cm}^2$
   b. Base 3.1 cm × altitude 3.7 cm = area $11.47^2 \approx 11\ \text{cm}^2$

6. a. Note that the altitude does not have to be drawn from a vertex; it just needs to be perpendicular to the base and to extend from the base to the opposite side:

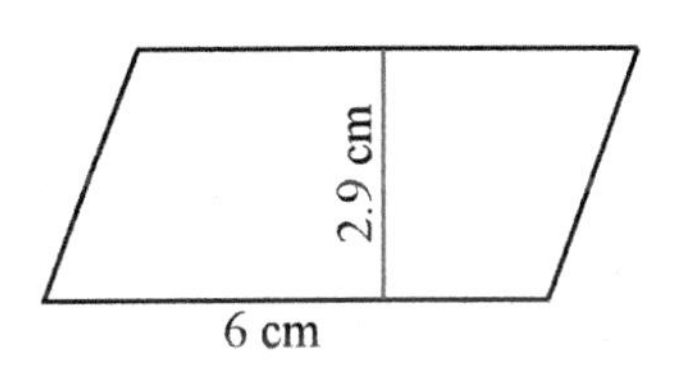

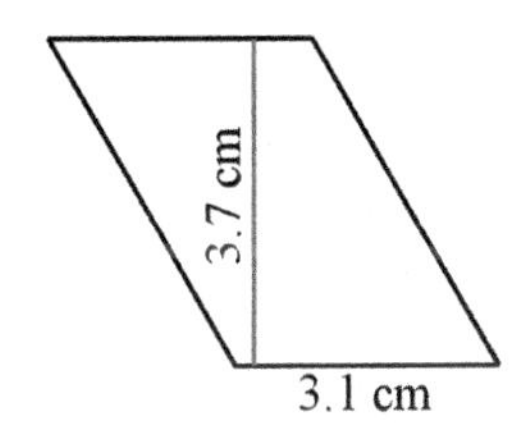

   b. (Left figure:)
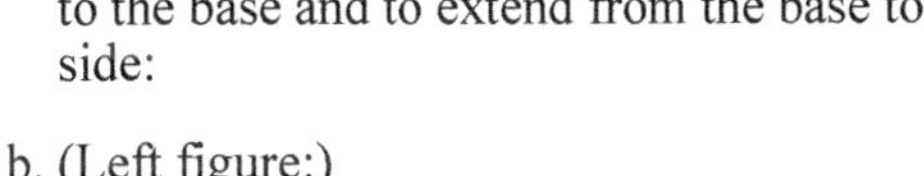
   Base 6.0 cm × altitude 2.9 cm = area 17.4 $\text{cm}^2 \approx 17\ \text{cm}^2$
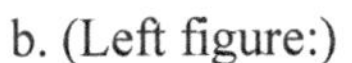
   (Right figure:)
   Base 3.1 cm × altitude 3.7 cm = area 11.47 $\text{cm}^2 \approx 11\ \text{cm}^2$

7. 156 sq. inches

8. 4.5 sq. meters

Puzzle corner:
The altitude of the larger parallelogram is half its base, so its area is 20 cm × 10 cm = 200 cm^2. The base and altitude of the smaller parallelogram are half the base and altitude of the larger one, so its area is (½ × 20 cm) × (½ × 10 cm) = 10 cm × 5 cm = 50 cm^2. The shaded area is the difference: 200 cm^2 – 50 cm^2 = 150 cm^2.

Area of Triangles, p. 113

1. $5 \times 4 / 2 = 10$ square units

2.

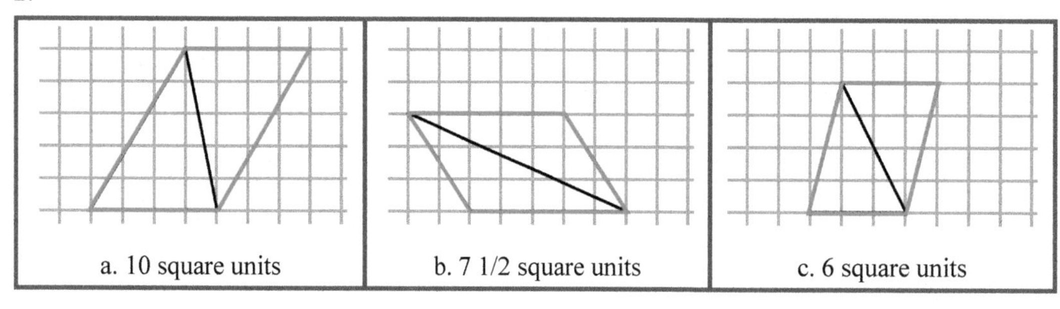

a. 10 square units | b. 7 1/2 square units | c. 6 square units

3.

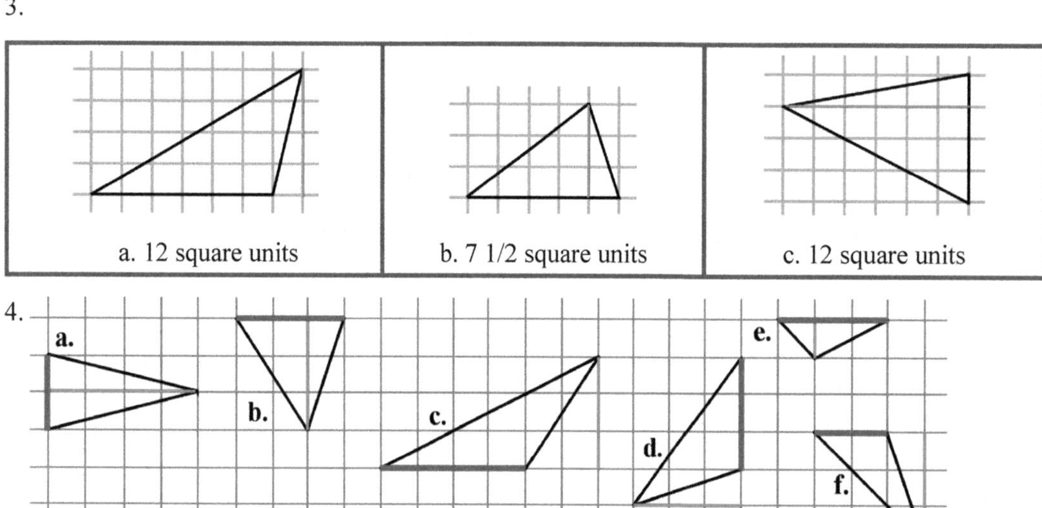

a. 12 square units | b. 7 1/2 square units | c. 12 square units

4.

a. 4 square units
b. 4 1/2 square units
c. 6 square units
d. 4 1/2 square units
e. 1 1/2 square units
f. 3 square units

5. a hexagon; area 3 + 12 + 3 = 18 square units

6. It is impossible to draw every triangle with an area of 12 square units on the grid because there is an infinite number. First, choose the length of the base and the altitude so that half of their product is 12: for example 6 and 4, 3 and 8, 2 and 12, 4.8 and 5, *etc*. Then, for each base and altitude, there exists an infinite variety of triangles, each with a different amount of "slant." For example:

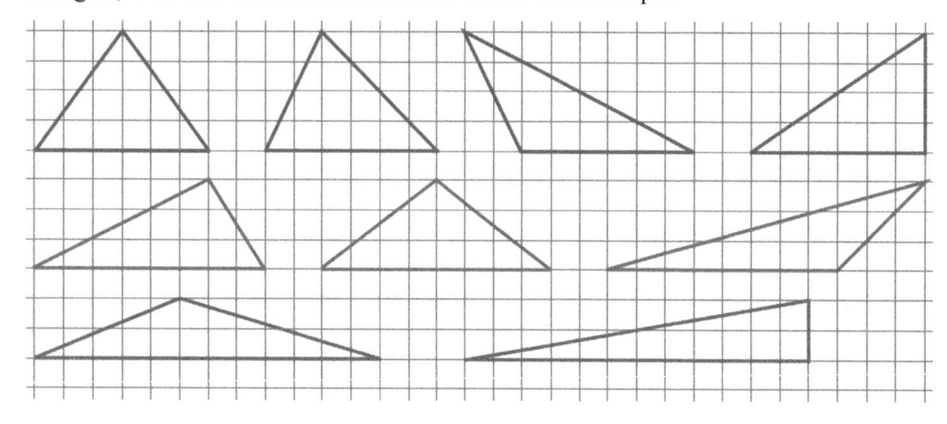

Area of Triangles, cont.

7\. a. 13 cm^2. Answers may vary slightly due to variations in measuring. The base and altitude may be chosen in three different ways. The figure shows one possibility, where the base is 6.8 cm, the altitude is 3.8 cm, and the area is 12.92 $cm^2 \approx 13$ cm^2.

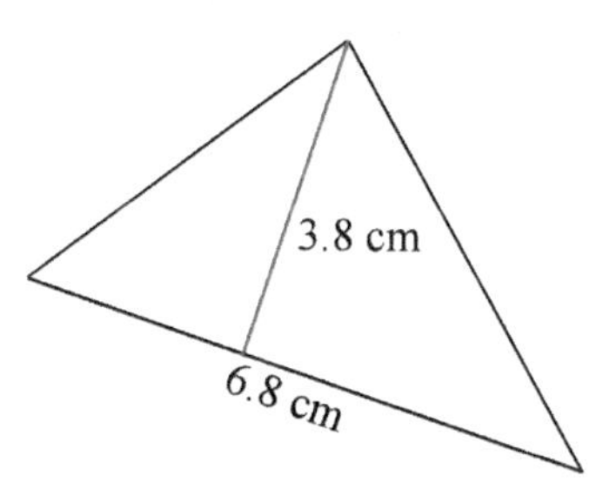

b. 2 1/8 sq. in. Again, answers may vary slightly due to variations in measuring. The base and altitude may be chosen in three different ways. So, the figure shows one possibility where the base is 4 in and the altitude is 1 1/16 in. The area is then 4 in × (1 1/16 in) ÷ 2 = 2 2/6 sq. in. = 2 1/8 sq. in.

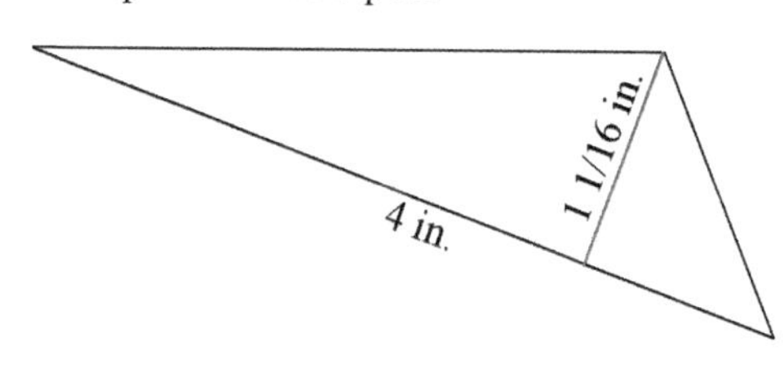

8\. Answers will vary; check the student's answers.

9\. Any triangle where the base times the altitude is 6 square inches will have an area of 3 square inches. For example, it could have a 3-inch base and a 2-inch altitude, or vice versa. Or a 6-inch base and a 1-inch altitude. Or a 4-inch base and a 1 1/2-in altitude. There are many possibilities.
Here is one example triangle: →

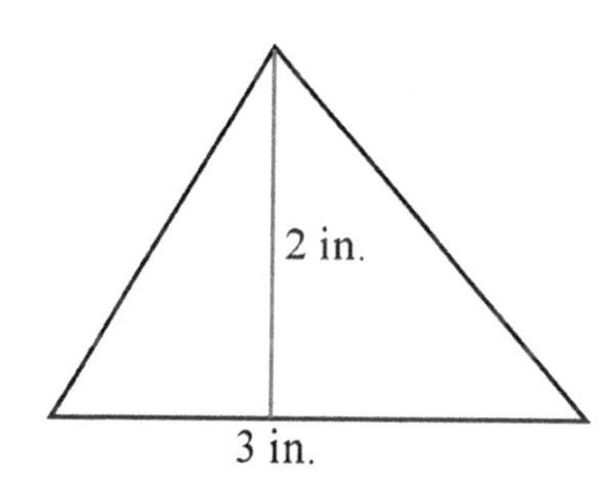

10\. Answers will vary; check the student's answers.

Area of Polygons, p. 116

1\. This figure is called a pentagon. Let's call the three triangles, 1, 2, and 3:
The area of triangle 1 is (2 units × 3 units) / 2 = 3 square units.
The area of triangle 2 is (2 units × 6 units) / 2 = 6 square units.
The area of triangle 3 is (3 units × 6 units) / 2 = 9 square units.
In total, the area of the pentagon is 3 + 6 + 9 = 18 square units.

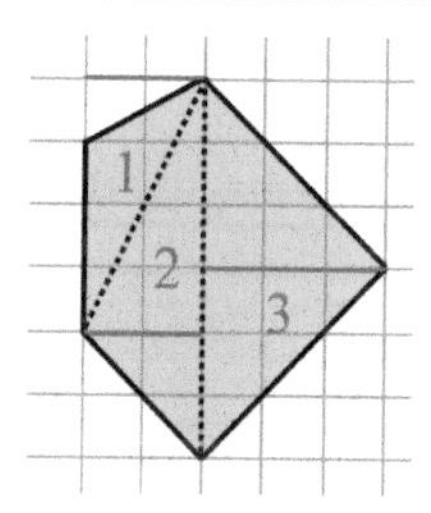

2\. The area of the enclosing rectangle is 5 × 6 = 30 square units. The small shaded triangles have areas of (2 × 1) ÷ 2 = 1, (3 × 3) ÷ 2 = 4 1/2, (3 × 3) ÷ 2 = 4 1/2, and (2 × 2) ÷ 2 = 2 square units, for a total of 12 square units. The area of our pentagon is the area of the rectangle minus the area of the shaded triangles, or 30 − 12 = 18 square units, the same answer that we got in Exercise #1.

3\. a. 15 − (3 + 2.5 + 3) = 6.5 square units
b. 20 − (2 + 3 + 1 + 4) = 10 square units

4\. a. The image shows the dog-polygon divided into triangles and rectangles. Note that the "paws" are 3/4 of a one square unit. (Why?) In total we get an area of 3/2 + 1 + 4 + 6 + 1/2 + 3 + 3/4 + 3/4 = 17 1/2 square inches.

b. 17 1/2 × 4 = 70 square inches. It may come as a surprise that the area is not doubled, but *quadrupled*. Consider one square in the image. If the side of the square is 1 inch, then its area is 1 square inch. But if the side of the square is 2 inches, then its area is 4 square inches. That is why all the areas are multiplied by 4.

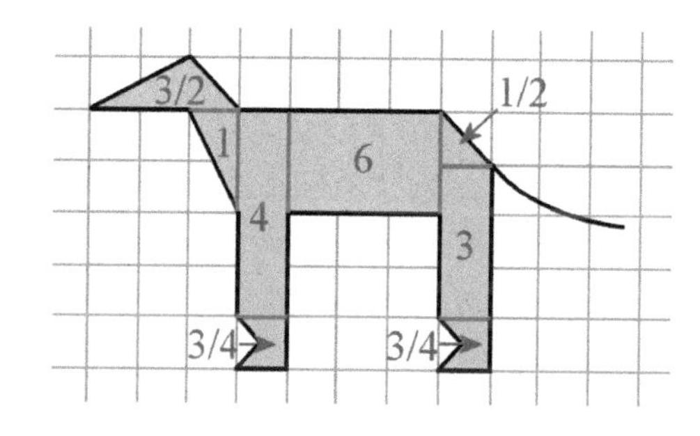

5\. a. 148 cm^2. The parallelogram with base 14 cm and altitude 8 cm has an area of 14 cm × 8 cm = 112 cm^2.
The triangle with a base of 9 cm and an altitude of 8 cm has an area of 9 cm × 8 cm ÷ 2 = 36 cm^2.
In total, the area of the figure is 112 + 36 = 148 cm^2.

b. 145 1/4 sq. in. The area of the rectangle is 7 in × 14 1/2 in = 7 in × 29/2 in = 203/2 in^2 = 101 1/2 in^2.
The area of the triangle is 7 in. × 12 1/2 in ÷ 2 = 7 in × 25/2 in ÷ 2 = 175/2 in^2 ÷ 2 = 175/4 in^2 = 43 3/4 in^2.

c. 21.4 sq. ft. The area of the rectangle is 6 ft × 3 ft = 18 sq. ft. The area of one of the triangles is 1.7 ft × 2 ft ÷ 2, so the area of both of them is 1.7 ft × 2 ft = 3.4 sq. ft. In total, the area is 21.4 sq. ft.

6\. Answers may vary, as there are two ways to divide the quadrilateral into triangles and also three ways to choose the base and altitude for each triangle. Moreover, the measuring may also vary a little.

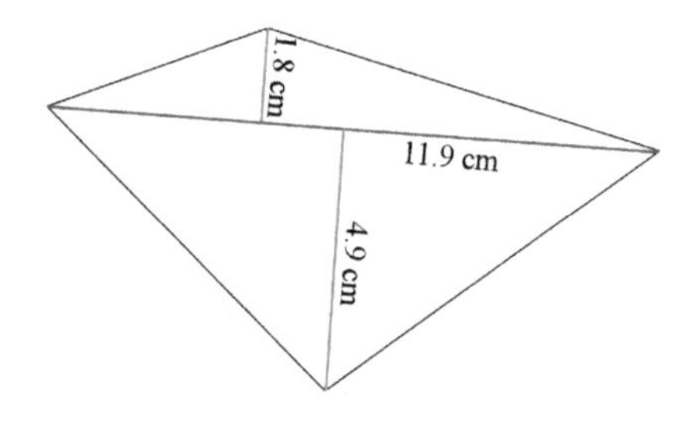

(11.9 cm × 1.8 cm) ÷ 2 + (11.9 cm × 4.9 cm) ÷ 2 = 39.865 cm^2 ≈ 40 cm^2.
(By measuring slightly differently the student might instead get 39 cm^2.)

Puzzle corner. Because of slight variations in measuring, the student's results may differ from these a little, or if the page wasn't printed at 100%, they could differ by a lot. Check the student's work.

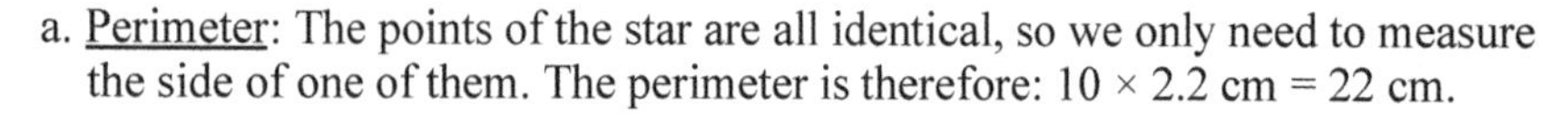

a. Perimeter: The points of the star are all identical, so we only need to measure the side of one of them. The perimeter is therefore: 10 × 2.2 cm = 22 cm.

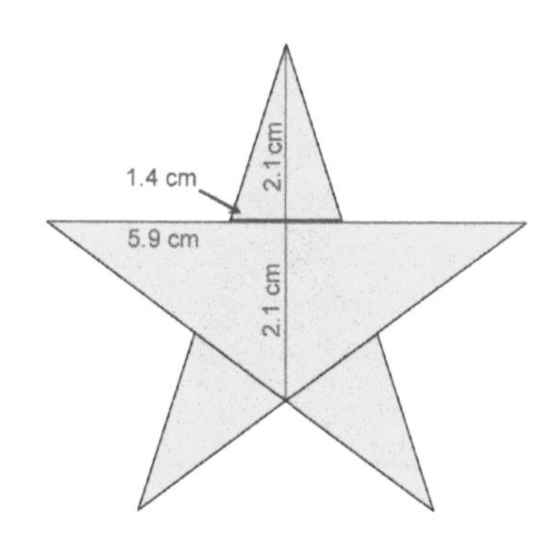

b. Area: You can divide the star into triangles in several ways. The picture shows one way, where we make one large triangle and three smaller ones. The altitude of the larger one is 2.1 cm, and its base is 5.9 cm. The three smaller triangles are identical and symmetrical, so we need to measure one side and altitude of one of them: The altitude is 2.1 cm and the base is 1.4 cm.

So the total area is: 5.9 cm × 2.1 cm ÷ 2 + 3 × (1.4 cm × 2.1 cm ÷ 2)
= 6.195 cm^2 + 3 × 1.47 cm^2 = 10.605 cm^2 ≈ 10.6 cm^2.

Polygons in the Coordinate Grid, p. 119

1\. a. A parallelogram.

b. See the image on the right.

c. 4 square units

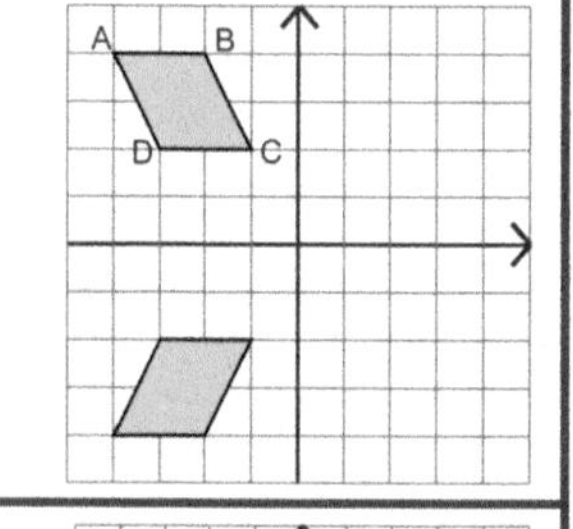

d. It is an isosceles right triangle.

e. See the image on the right.

f. 4 square units

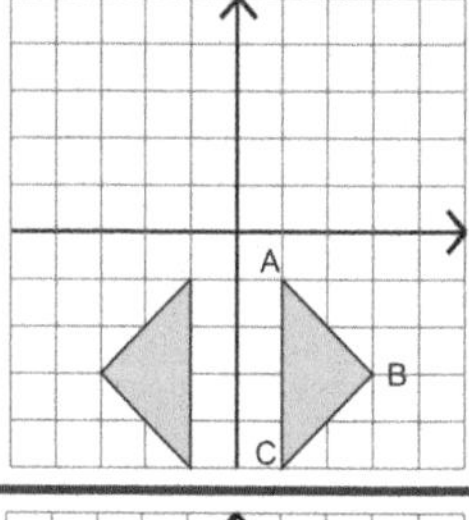

g. A hexagon.

h. See the image on the right.

i. 4 square units

j. and k. Check the student's answers.

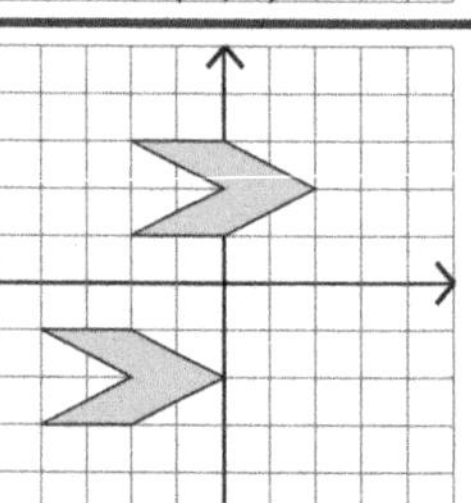

Polygons in the Coordinate Grid, cont.

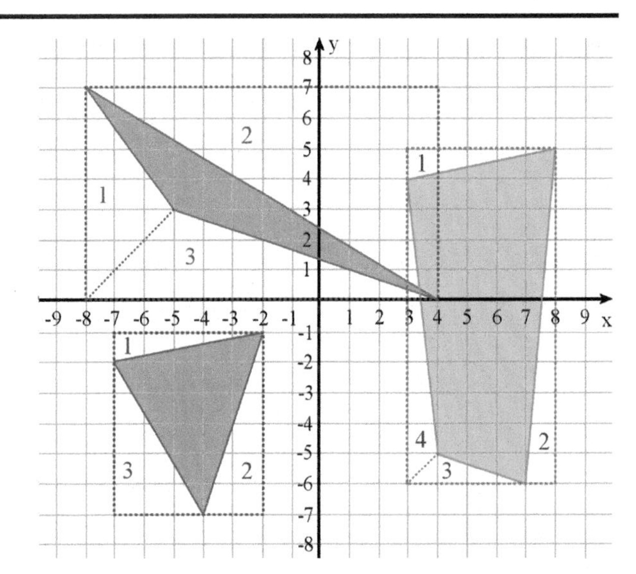

2\. a. 13.5 square units.
The area of the outer rectangle is $12 \times 7 = 84$ square units.
The area of triangle 1 is $7 \times 3 \div 2 = 10.5$ square units.
The area of triangle 2 is $12 \times 7 \div 2 = 42$ square units.
The area of triangle 3 is $12 \times 3 \div 2 = 18$ square units.
Therefore, the area of the colored triangle is
$84 - 10.5 - 42 - 18 = 13.5$ square units.

b. 14 square units.
The area of the outer rectangle is $5 \times 6 = 30$ square units.
The area of triangle 1 is $5 \times 1 \div 2 = 2.5$ square units.
The area of triangle 2 is $2 \times 6 \div 2 = 6$ square units.
The area of triangle 3 is $3 \times 5 \div 2 = 7.5$ square units.
Therefore, the area of the colored triangle is
$30 - 2.5 - 6 - 7.5 = 14$ square units.

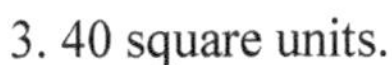

3\. 40 square units.
The area of the outer rectangle is $5 \times 11 = 55$ square units.
The area of triangle 1 is $5 \times 1 \div 2 = 2.5$ square units.
The area of triangle 2 is $1 \times 11 \div 2 = 5.5$ square units.
The area of triangle 3 is $4 \times 1 \div 2 = 2$ square units.
The area of triangle 4 is $10 \times 1 \div 2 = 5$ square units.
Therefore, the area of the quadrilateral is $55 - 2.5 - 5.5 - 2 - 5 = 40$ square units.

4\. Each side is 2 units long. The perimeter is 16 units.

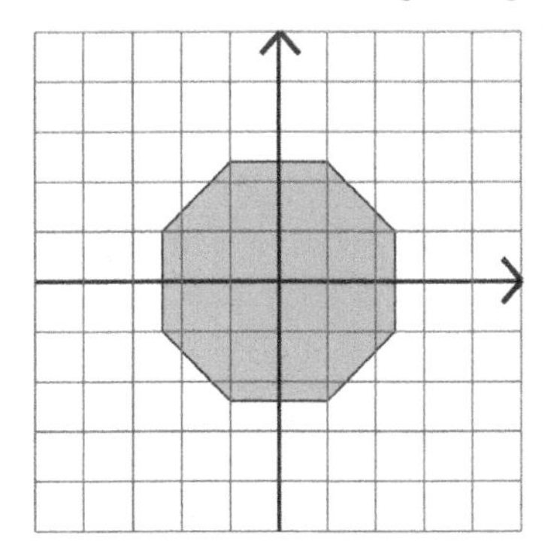

5\. Letter N. To calculate its area, we can divide it into two identical rectangles and a parallelogram. Each rectangle has an area of $15 \times 80 = 1{,}200$ square units. The area of the parallelogram is $25 \times 40 = 1{,}000$ square units. The total area is 3,400 square units.

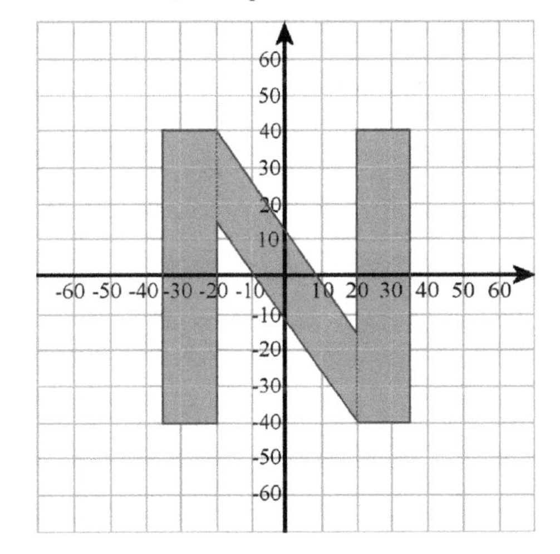

Polygons in the Coordinate Grid, cont.

6. For example, here are three possibilities:

Pool Design 1:	Pool Design 2:	Pool Design 3:
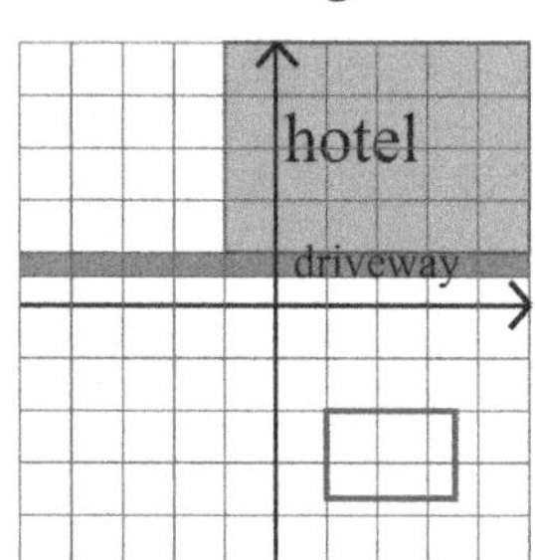	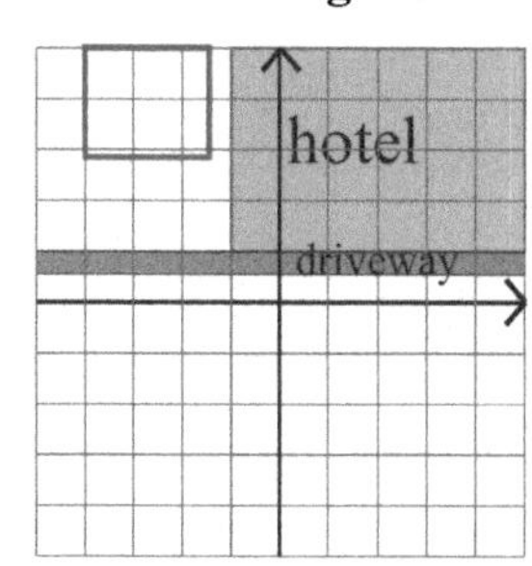	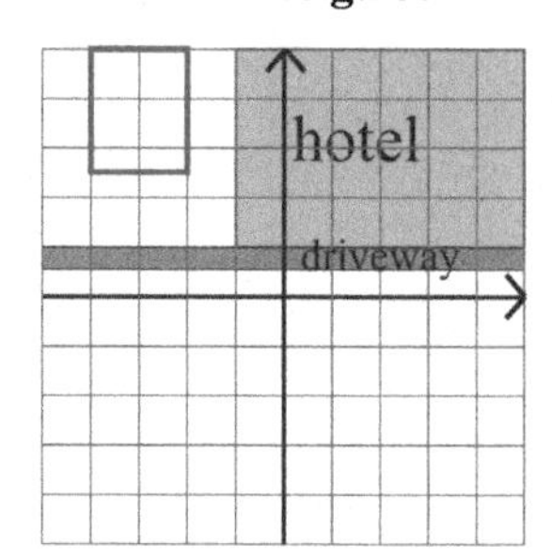
Coordinates of the corners:	Coordinates of the corners:	Coordinates of the corners:
(5, -10), (17.5, -10), (17.5, -18), (5, -18)	(-20, 14), (-20, 25), (-7.5, 25), (-7.5, 14)	(-20, 12.5), (-20, 25), (-10, 25), (-10, 12.5)
Area: 12.5 m × 8 m = 100 m^2	Area: 12.5 m × 11 m = 137.5 m^2	Area: 10 m × 12.5 m = 125 m^2
Distance to the driveway: 12 1/2 m	Distance to the driveway: 9 m	Distance to the driveway: 7.5 m

Area and Perimeter Problems, p. 122

1. We multiply the two dimensions by 40, and then convert the results into feet:
   40 × 9 in = 360 in = 30 ft, and 40 × 12 in = 480 in = 40 ft. The house is 30 ft × 40 ft.
   Its area is 1,200 square feet, and its perimeter is 140 ft.

2. Divide the house into three rectangles. This can be done in different ways; one way is on the right.
   Rectangle 1: 42 ft × 15 ft = 630 sq. ft. Rectangle 2: 40 ft × 30 ft = 1,200 sq. ft.
   Rectangle 3: 15 ft × 10 ft = 150 sq. ft.
   In total, the area is 630 sq. ft. + 1,200 sq. ft. + 150 sq. ft.. = 1,980 sq. ft.

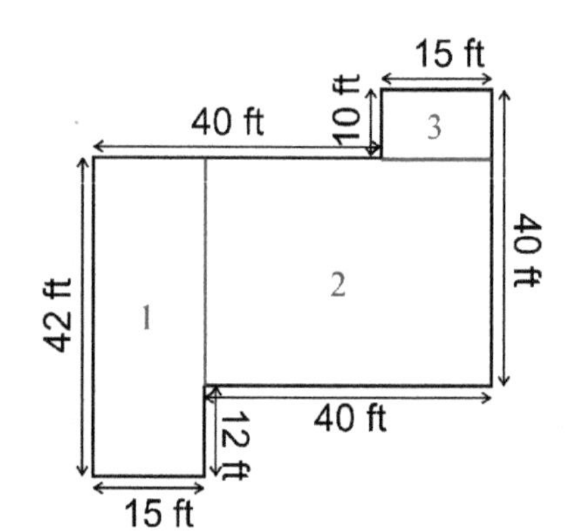

3. a. The area of the house is 80 m^2, and the area of the yard is 320 m^2.
   So, the house takes up 80/320 = <u>1/4 of the area of the yard</u>.

   b. 320 m^2 − 80 m^2 = <u>240 m^2</u>.

   c. The perimeter without the gate is 20 m + 16 m + 20 m + 16 m − 3m = 69 m.
   The cost is 69 × \$11.59 + \$120 = <u>\$919.71</u>.

4. a. 150 ft × 150 ft = 22,500 sq. ft.
   b. To find the area, we divide the trapezoid of part 1 into a right triangle and a rectangle:
   The triangle has an area of 100 ft × 150 ft ÷ 2 = 7,500 sq. ft. The rectangle has an area of 150 ft × 75 ft = 11,250 sq. ft. The total area of part 1 is 18,750 sq. ft.
   c. The total area is 41,250 sq. ft.
   d. 45%, or to be more exact, 45.4545%
   e. 55%, or to be more exact, 54.5455%
   f. \$20,250 and 24,750 using the rounded percentages, or \$20,455 and \$24,545 using the more exact percentages.

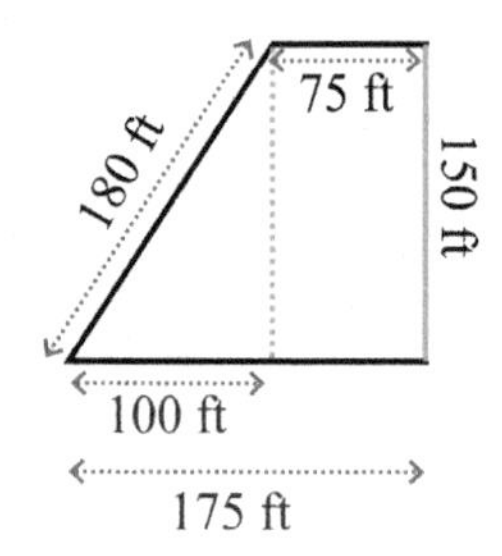

5. a. The darker areas consist of 8 half-squares, with a total area of 4 square units.
   The total area is 16. Therefore, the darker part takes up 1/4 of the design.
   b. The darker areas consist of 6 squares. The total area is 12.
   So, the darker part takes up 1/2 of the design.

6. The area of the figure is the sum of the four areas marked 1, 2, 3, and 4:
   Triangles 1 and 2: (3.5 ft × 6 ft) ÷ 2 = 10.5 ft^2.
   Rectangle 3: 10 ft × 6 ft = 60 ft^2.
   Rectangle 4: (10 + 3.5 + 3.5) ft × 7 ft = 17 ft × 7 ft = 119 ft^2.
   The total area is the sum: 10.5 ft^2 + 10.5 ft^2 + 60 ft^2 + 119 ft^2 = 200 ft^2.

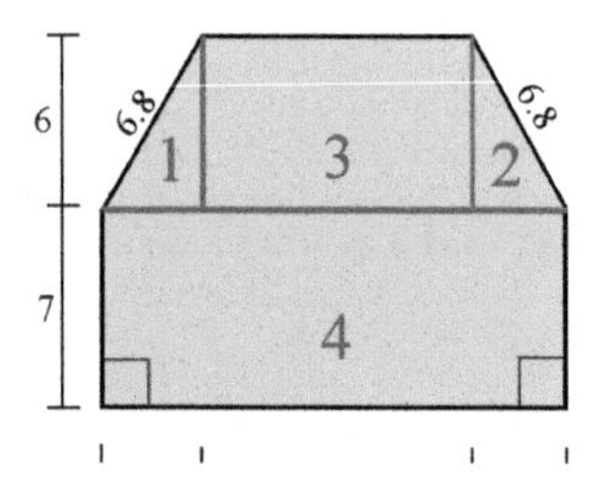

Nets and Surface Area 1, p. 124

1. a. no b. no c. no
   d. yes e. no f. yes

2. 6 faces × 4 cm × 4 cm = 96 cm^2

3. a. (2) b. (1) or (3) c. (4) d. (3) or (1)

4. Did you notice that there are as many units of volume as there are little cubes in each solid and as many units of area as there are squares in each net?

   a. A = 14 cm^2 V = 3 cm^3 b. A = 22 cm^2 V = 6 cm^3 c. A = 32 cm^2 V = 12 cm^3 d. A = 22 cm^2 V = 6 cm^3

5. Notice that the surface area quadruples, and the volume increases 8-fold.

   a. A = 56 cm^2 V = 24 cm^3 b. A = 88 cm^2 V = 48 cm^3 c. A = 128 cm^2 V = 96 cm^3 d. A = 88 cm^2 V = 48 cm^3

6. a. One face has an area of 96 sq. in ÷ 6 = 16 square inches.
   b. 4 inches
   c. 4 in × 4 in × 4 in = 64 cubic inches

7. One edge is 3 feet (3 × 3 × 3 = 27).
   So the surface area is:
   6 faces × 3 feet × 3 feet = 54 square feet.

8. a. rectangular prism
   b. cube
   c. triangular prism

9. a. solid: square pyramid
   surface area:
   15 cm × 15 cm + 4 × (15 cm × 13 cm ÷ 2)
   = 225 cm^2 + 390 cm^2 = 615 cm^2.

   b. solid: rectangular prism
   surface area: 2 × 14 in × 7 in
   + 2 × 14 in × 11 in + 2 × 7 in × 11 in
   = 196 sq. in. + 308 sq. in. + 154 sq. in.
   = 658 sq. in.

Nets and Surface Area 2, p. 127

1. Answers may vary. Check the student's work.
   For example:

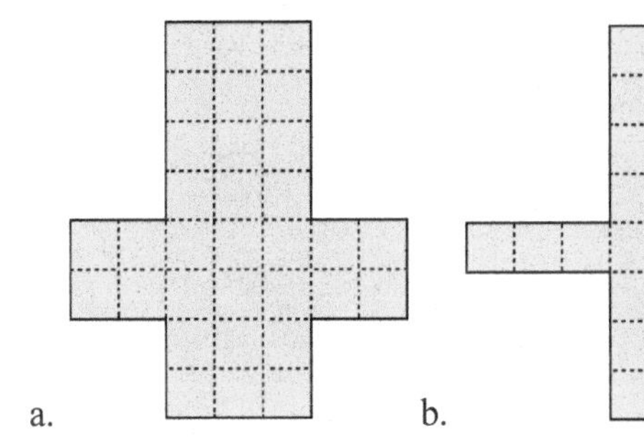

a. b.

2. expression (1)

3.

Top and bottom:
2 × 6 cm × 8 m = 2 × 48 cm^2 = 96 cm^2
Back and front:
2 × 6 cm × 7.5 cm = 90 cm^2
The two sides:
2 × 8 cm × 7.5 cm = 120 cm^2
Total: 96 cm^2 + 90 cm^2 + 120 cm^2 = 306 cm^2

4. a.

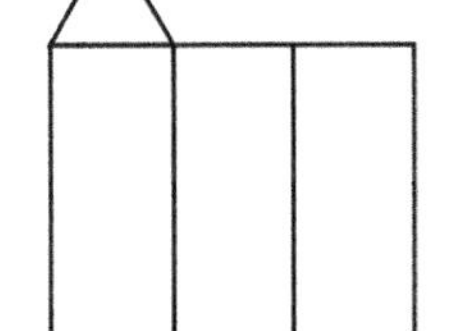

or

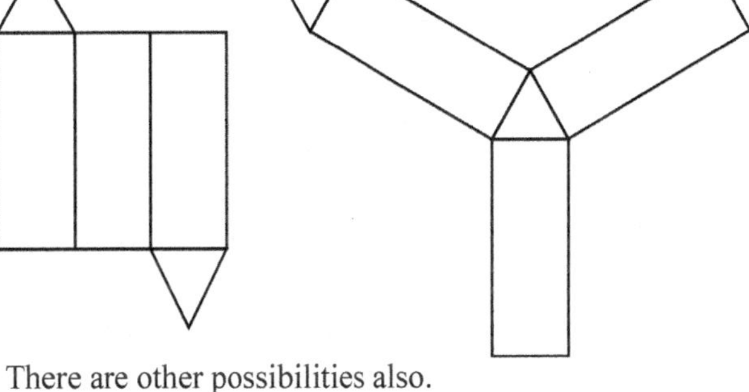

There are other possibilities also.

   b. The two base triangles have a total area of
   2 × 4 cm × 4.5 cm ÷ 2 = 18 cm^2.
   The other three faces are rectangles, with a total area of 3 × 4 cm × 11 cm = 132 cm^2.
   In total, the surface area is <u>150 cm^2</u>.

5. 6 × 20 cm × 20 cm = 2,400 cm^2.

6. a. The width is 20 cm. The length is 30 cm + 24 cm + 30 cm + 24 cm = 108 cm.

   b. The area is 108 cm × 20 cm =2,160.

7. a.

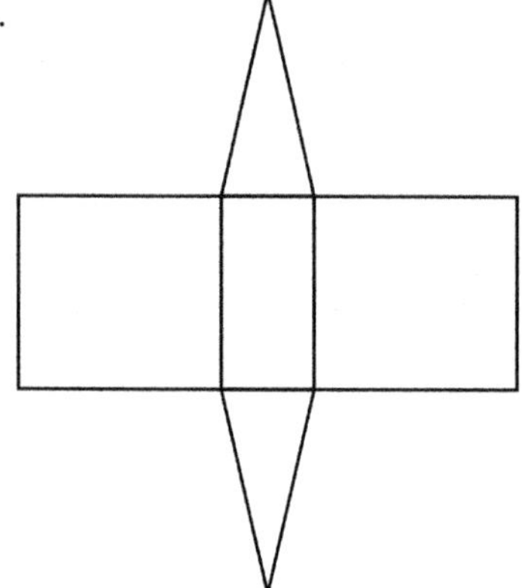

or

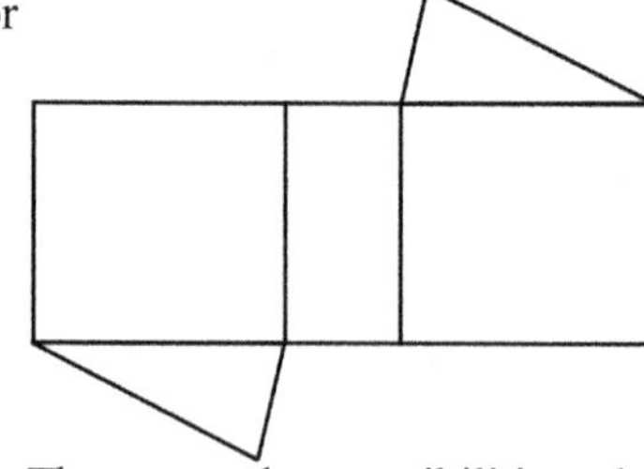

There are other possibilities, also.

Surface area:
Bottom: 6 m × 12 m = 72 m^2.
End triangles: 2 × 6 m × 12.6 m ÷ 2 = 75.6 m^2.
The two other faces: 2 × 12 m × 13 m = 312 m^2.
Total surface area is 459.6 m^2.

7. b.

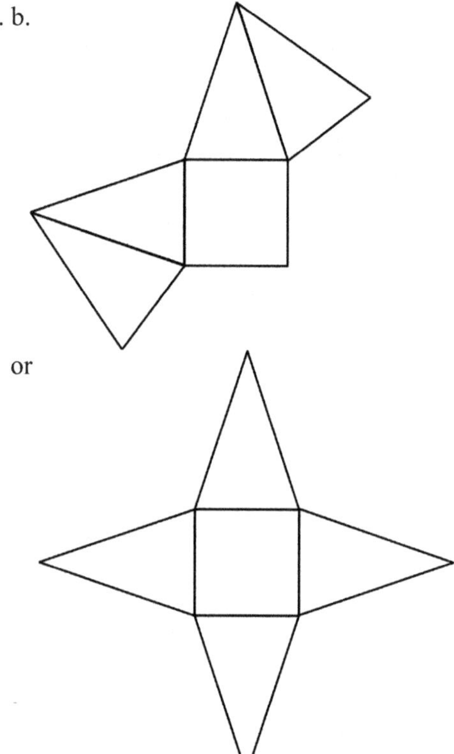

or

There are other possibilities, also.

Surface area: Bottom: 10 in × 10 in = 100 sq. in.
Triangular faces: 4 × 10 in × 15 in ÷ 2 = 300 sq. in.
Total: 400 sq. in.

8. a.

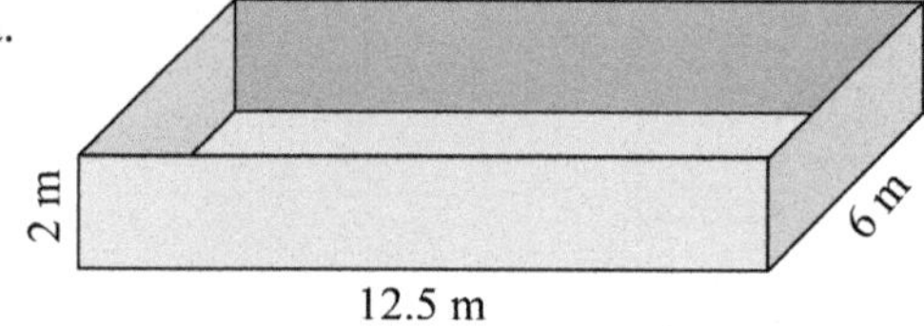

b. 2 × 12.5 m × 2 m + 12.5 m × 6 m + 2 × 2 m × 6 m
= 50 m^2 + 75 m^2 + 24 m^2 = 149 m^2.

c. $1,475.10

Converting Between Area Units, p. 131

Teaching box:

1 sq. ft = 12 in × 12 in = 144 sq. in.
1 cm^2 = 10 mm × 10 mm = 100 mm^2

1. 1 sq. yd. = 3 ft × 3 ft = 9 sq. ft.

2. Each side is 100 cm.
The area in square *meters* is 1 m × 1 m = 1 m^2.

The area in square *centimeters* is
100 cm × 100 cm = 10,000 cm^2.

3. The area in square *meters* is 2 m × 2 m = 4 m^2.

The area in square *centimeters* is
200 cm × 200 cm = 40,000 cm^2.

4. The area in square *feet* is 3 ft × 8 ft = 24 ft^2.

The area in square *inches* is 36 in × 96 in = 3,456 in^2.

5. a. 1,760 yd × 1,760 yd = 3,097,600 sq. yd.

b. 1,000 m × 1,000 m = 1,000,000 m^2.

6. Answers may vary if the file is not printed at 100%, or because of slight inaccuracies and variations in measuring.

a. 600 mm^2; 6 cm^2
b. 480 mm^2; 4.8 cm^2. (It is a parallelogram with base 3.2 cm and height 1.5 cm.)
c. 273 mm^2; 2.73 cm^2. Note that you can draw an altitude for a triangle in 3 different ways; one is shown below.

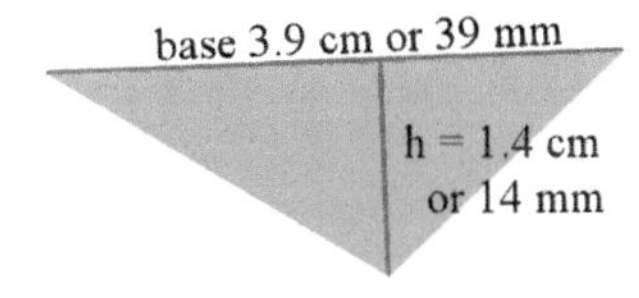

7. Multiply by 100. You will get 5,800 square millimeters.

8. a. 0.8 mi × 2 mi = 1.6 sq. mi.

b. (0.8 × 5,280 ft) × (2 × 5,280 ft)
= 44,605,440 square feet

9. 0.2 km is 200 m, and 0.15 km is 150 m.
The area is 200 m × 150 m = 30,000 m^2.

10. Answers may vary if the file is not printed at 100% or because of slight variations in measuring.

a. Divide the quadrilateral into two triangles. Then draw the altitudes for the two triangles. Then, measure the bases and the altitudes for both triangles. For example:

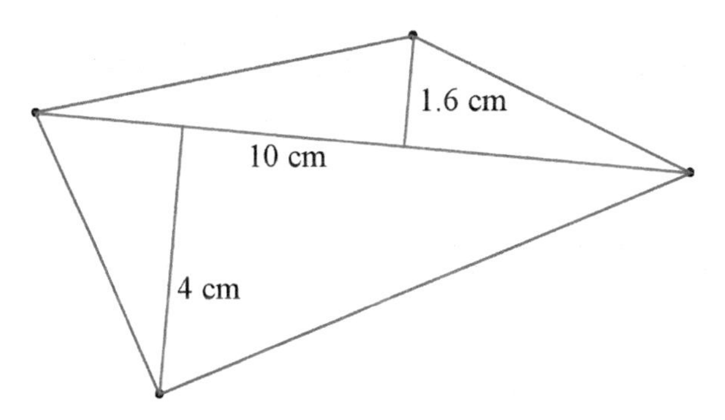

The area is
100 mm × 40 mm ÷ 2 + 100 mm × 16 mm ÷ 2
= 2,000 mm^2 + 800 mm^2 = 2,800 mm^2.

b. The area is
10 cm × 4 cm ÷ 2 + 10 cm × 1.6 cm ÷ 2
= 20 cm^2 + 8 cm^2 = 28 cm^2.

Volume of a Rectangular Prism with Sides of Fractional Length, p. 133

1.

a.	b. 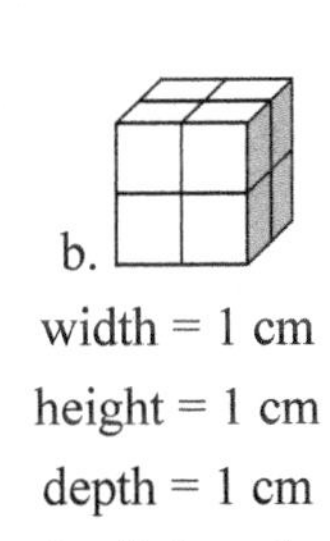 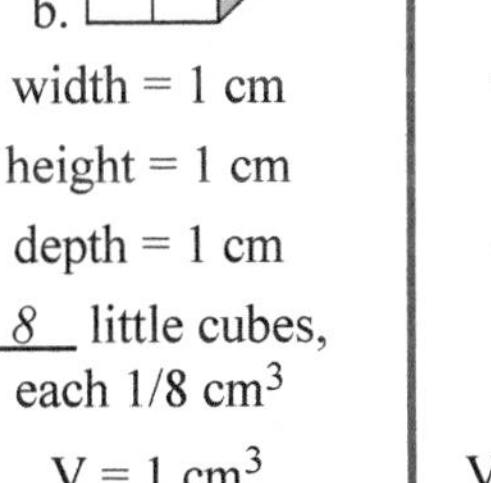	c. 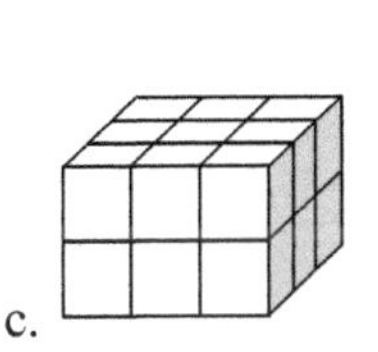	d. 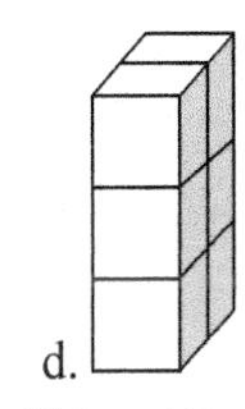
width = *1/2* cm	width = 1 cm	width = 1 1/2 cm	width = 1/2 cm
height = 1/2 cm	height = 1 cm	height = 1 cm	height = 1 1/2 cm
depth = 1/2 cm	depth = 1 cm	depth = 1 1/2 cm	depth = 1 cm
1 little cube, 1/8 cm^3	*8* little cubes, each 1/8 cm^3	18 little cubes, each 1/8 cm^3	6 little cubes, each 1/8 cm^3
V = 1/8 cm^3	V = 1 cm^3	V = 9/4 or 2 1/4 cm^3	V = 6/8 or 3/4 cm^3

2.

a. V = 1 1/2 cm × 1 cm × 1 1/2 cm = 3/2 cm × 1 cm × 3/2 cm = 9/4 cm^3 = 2 1/4 cm^3	b. V = 1/2 cm × 1 1/2 cm × 1 cm = 1/2 cm × 3/2 cm × 1 cm = 3/4 cm^3

3. This time, the edges of each little cube measure 1/3 inch.

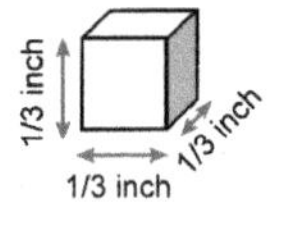

We put 27 of the little cubes together to form one cubic inch.

Since the big cube measures 1 cubic inch, and there are 27 little cubes,

the volume of each little cube is 1/27 cubic units. This is the same answer that we find

by multiplying: $V = \frac{1}{3}\text{ in} \times \frac{1}{3}\text{ in} \times \frac{1}{3}\text{ in} = \frac{1}{27}\text{ in}^3$.

4. (i)

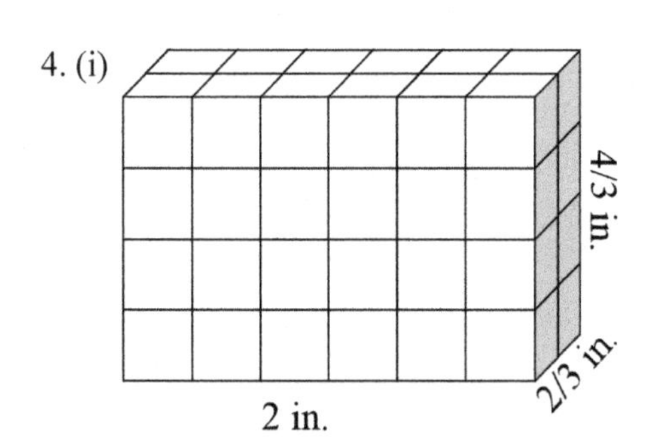

The model needs to be 4 by 6 by 2 blocks, but it can be in a different orientation from this picture.

(ii) There are 48 cubes.

(iii) One cube is 1/27 cubic inch. The total volume is 48 × (1/27) in^3 = 48/27 in^3 = 16/9 in^3, which is the same as you would get by multiplying the three dimensions, 4/3 in, 2 in, and 2/3 in.

5. (i)

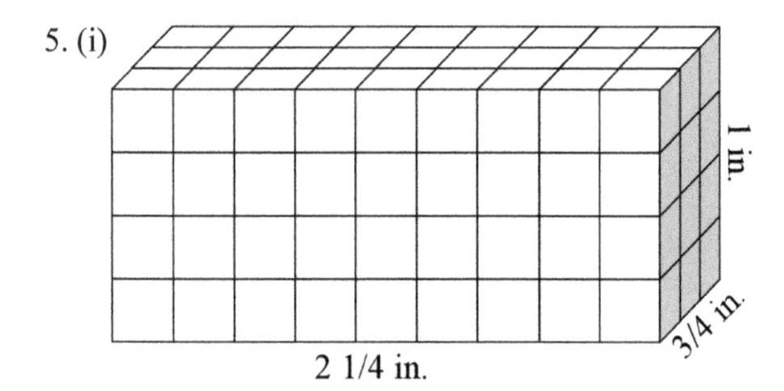

The model needs to be 9 by 4 by 3 blocks but can be in a different orientation from this picture.

(ii) There are 108 cubes.

(iii) One cube is 1/64 cubic inch. The total volume is 108 × (1/64) in^3 = 108/64 in^3 = 27/16 in^3 which is the same as you would get by multiplying the three dimensions, 3/4 in, 2 1/4 in, and 1 in.

6.

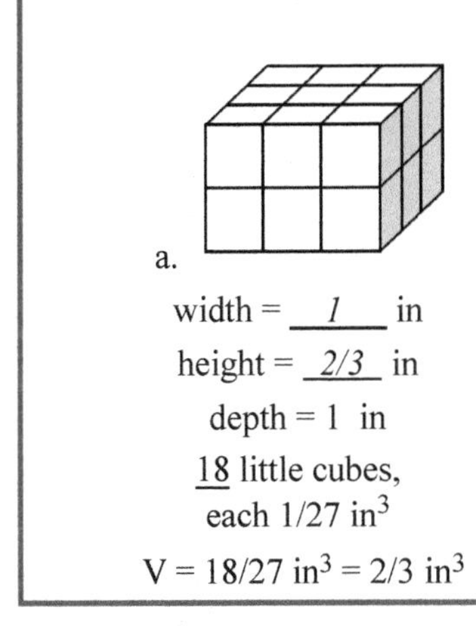

a.

width = *1* in
height = *2/3* in
depth = 1 in
18 little cubes, each $1/27\ in^3$
$V = 18/27\ in^3 = 2/3\ in^3$

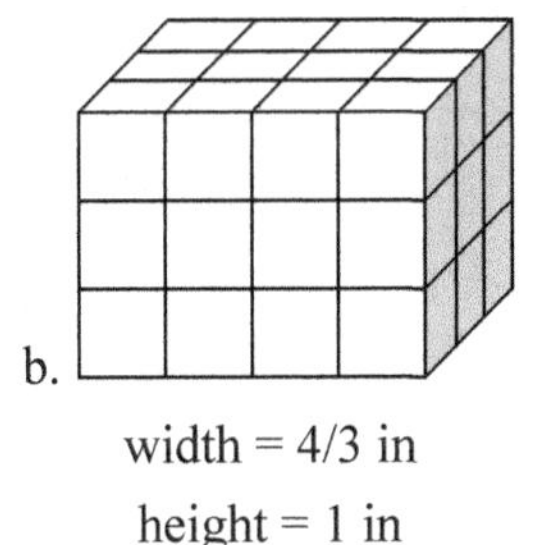
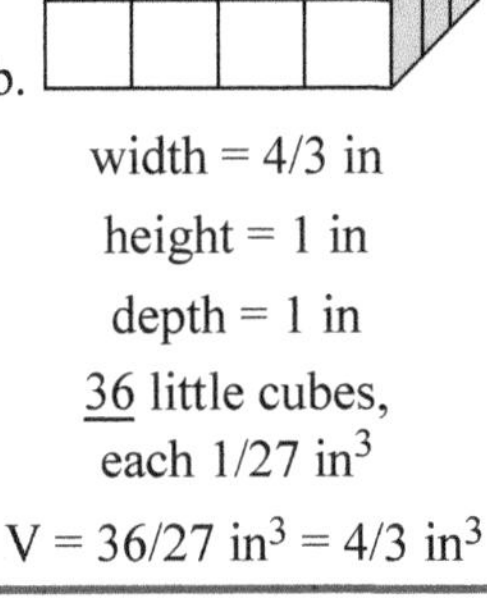

b.

width = 4/3 in
height = 1 in
depth = 1 in
36 little cubes, each $1/27\ in^3$
$V = 36/27\ in^3 = 4/3\ in^3$

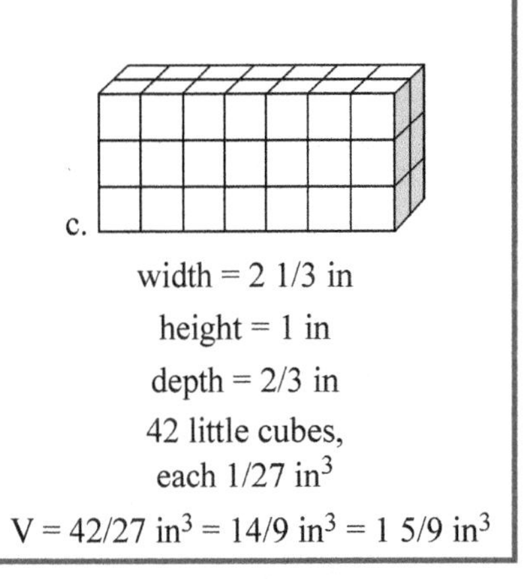

c.

width = 2 1/3 in
height = 1 in
depth = 2/3 in
42 little cubes, each $1/27\ in^3$
$V = 42/27\ in^3 = 14/9\ in^3 = 1\ 5/9\ in^3$

7. a. $V = 1\ in \times 2/3\ in \times 1\ in\ =\ 2/3\ in^3$

b. $V = 4/3\ in \times 1\ in \times 1\ in\ =\ 4/3\ in^3$

c. $V = 2\ 1/3\ in \times 1\ in \times 2/3\ in\ =\ 7/3\ in \times 1\ in \times 2/3\ in\ =\ 14/9\ in^3$

Volume Problems, p. 136

1. a. $7.2\ cm \times 7.2\ cm \times 18\ cm = 933.12\ cm^3 \approx 933\ cm^3$
   b. 933 ml
   c. 896 ml

2. a. In inches, the box is 36 in by 14 in by 16 in. In feet, it is 3 ft by 1 1/6 ft by 1 1/3 ft (or 3 ft by 1.167 ft by 1.333 ft).

   b. $3\ ft \times 1\ 1/6\ ft \times 1\ 1/3\ ft = 3\ ft \times 7/6\ ft \times 4/3\ ft\ = 28/6\ ft^3 = 14/3\ ft^3 = 4\ 2/3\ ft^3$

3. a. $2\ 1/4\ in \times 2\ 1/4\ in \times 2\ 1/4\ in = 9/4\ in \times 9/4\ in \times 9/4\ in = 729/64\ in^3 = 11\ 25/64\ in^3$.

   b. The box is 24 in by 12 in by 12 in You can pack 10 cubes into the 2-ft dimension because 10 × 2 1/4 in = 22 1/2 in. and one more would not fit. You can pack 5 cubes into each of the 1-ft dimensions because 5 × 2 1/4 in = 11 1/4 in. In total, you can pack 10 × 5 × 5 = 250 cubes in the box.

Puzzle corner. The table below lists both the exact amounts in gallons and liters, and the rounded amounts.

For example, you can calculate the volumes in cubic inches, then use the conversion factor 1 cubic inch = 0.554113 liquid ounces to change those into liquid ounces, and then divide by 128 to get gallons.

For liters, simply multiply the dimensions in centimeters to get cubic centimeters, then divide by 1000.

STANDARD TANKS					
DIMENSIONS	**VOLUME**		**DIMENSIONS**	**VOLUME**	
Inches	**Gallons / exact**	**Gallons**	**Centimeters**	**Liters**	**Liters**
16 × 8 × 10	5.54113	5.5	41 × 20¼ × 25½	21.171375	21
20 × 10 × 12	10.3896	10.5	51 × 25 × 30	38.25	38
24 × 12 × 16	19.9481	20.0	61 × 30½ ×40½	75.35025	75
36 × 12 × 17	31.7922	31.75	91½ × 30½ × 43	120.00225	120
36 × 18 × 16	44.8832	45.0	92½ × 46 × 41	174.455	174
36 × 18 × 18	50.4935	50.5	92½ × 46 × 46	195.73	196

Mixed Review, p. 138

1. They put 20 lb of flour into the cellar. Next, 3/8 of the remaining 40 lb is 15 lb. The neighbor got 15 lb.

2. a. 0.072 b. 1.54 c. 25,000 d. 0.0072 e. 0.2 f. 2.1

3. a. 7/10 < 8/10 < 5/6 < 7/8 < 9/10

   b. 11/10 < 9/8 < 7/6 < 12/10 < 10/8

4. a. 0.9 L = 9 dl = 90 cl = 900 ml

   b. 2,800 m = 2.8 km = 28,000 dm = 280,000 cm

   c. 56 g = 560 dg = 5,600 cg = 56,000 mg

5.

a. 76 oz = 4.75 lb	c. 3.6 gal = 14.4 qt	e. 2.67 mi = 14,098 ft
b. 98 in = 8.17 ft	d. 0.483 lb = 7.73 oz	f. 5.09 ft = 5 ft 1 in

6.

a. 134 kg	=	134 kg ·	$\frac{2.2 \text{ lb}}{1 \text{ kg}}$	= 294.8 lb
b. 156 in	=	156 in ·	$\frac{2.54 \text{ cm}}{1 \text{ in}}$	= 396.24 cm

7.

a.		b.		c.	
$0.2m = 6$	**÷ 0.2**	$0.3p = 0.09$	**÷ 0.3**	$y - 1.077 = 0.08$	**+ 1.077**
$m = 6 \div 0.2$		$p = 0.09 \div 0.3$		$y = 0.08 + 1.077$	
$m = 30$		$p = 0.3$		$y = 1.157$	

8. a. See the picture on the right.
   b. The unit rates are:

   2 1/2 squares for 1 triangle

   2/5 triangles for 1 square

9.

a. $5 + (-8) = -3$	b. $-11 + (-9) = -20$	c. $2 + (-17) = -15$	d. $2 - (-8) = 10$
$(-5) + 8 = 3$	$9 - 11 = -2$	$-3 - 8 = -11$	$-8 - (-2) = -6$

10. Movements 1. and 2. are shown in the picture. The last movement (the reflection) results in a figure that overlaps the previous one so there is no arrow to show it in the picture.

    The coordinates after the transformations are (−4, 0), (3, 0), (1, −2), and (−3, −2).

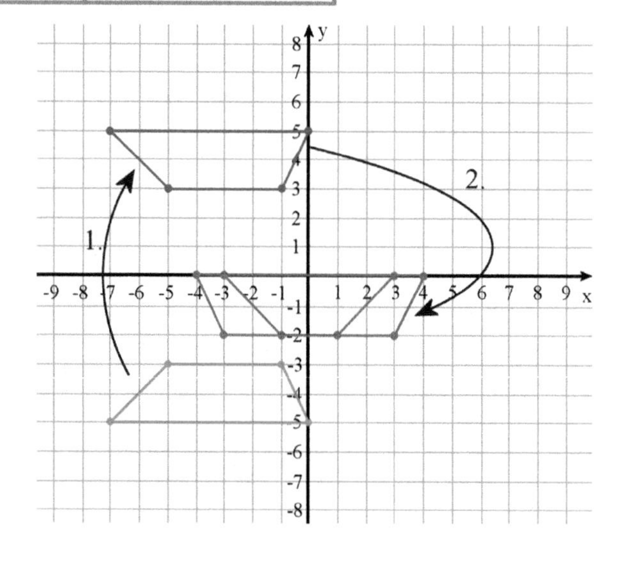

Mixed Review, cont.

11. The area is $8 \times 11 \div 2 = 44$ square units

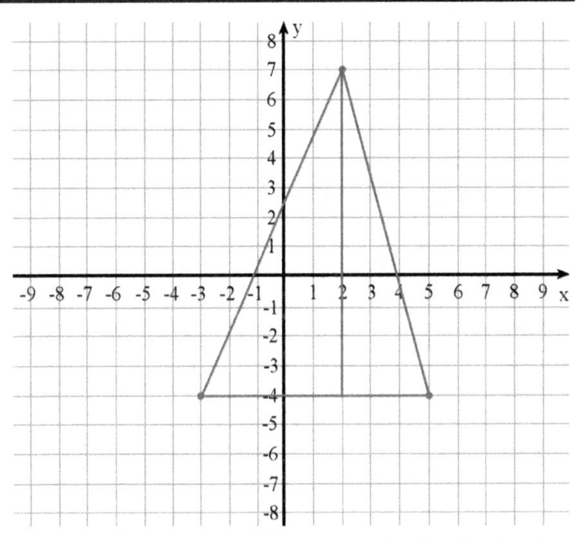

12. a. You can choose any two letters, but a good choice is *t* for time and *d* for distance.

b.

time (hours)	0	1	2	3	4	5	6	7	8	9
distance (meters)	0	4	8	12	16	20	24	28	32	36

c. $d = 4t$ or $t = d/4$

d. Because the length of the tunnel depends on the time that the mole has been digging, the distance (d) is the dependent variable, and the time (t) is the independent variable.

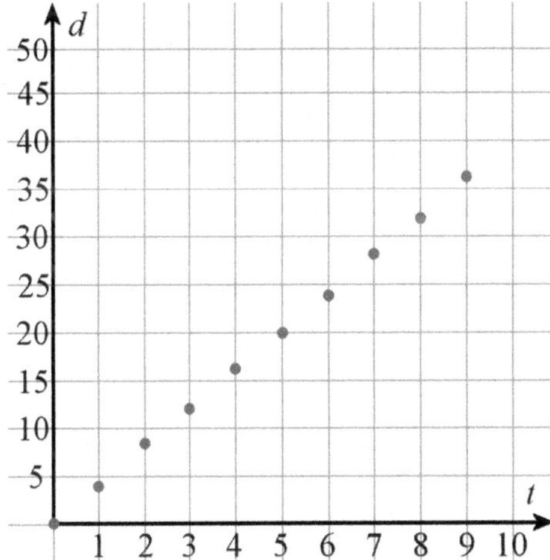

13. a. Dividing a number by 5 is the same as multiplying it by 1/5. Example: $8 \div 5$ is the same as $8 \times 1/5$. Both equal 8/5. Another example: $20 \div 5$ is equal to $20 \times 1/5$. Both are equal to 4. (Examples will vary. Check the student's work.)

b. Dividing a number by 2/3 is the same as multiplying it by 3/2.
Example: $6 \div (2/3)$ is the same as $6 \times (3/2)$. Both equal 9. (Examples will vary. Check the student's work.)

14. a. 5/8 = 63% (exactly 62.5%)
b. 6/25 = 24%

15.

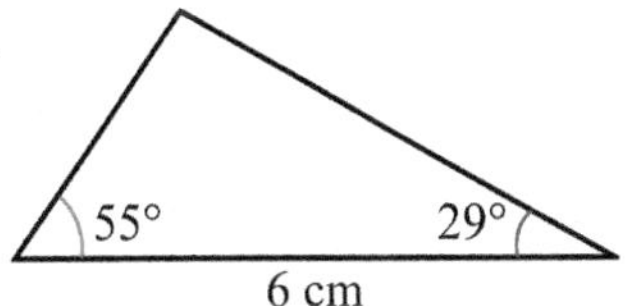

16. First, draw the 66° angle, then measure the two 7.5 cm sides. Or, draw one of the 7.5 cm sides first. Then you can use the compass for the last two sides. The image is not to scale.

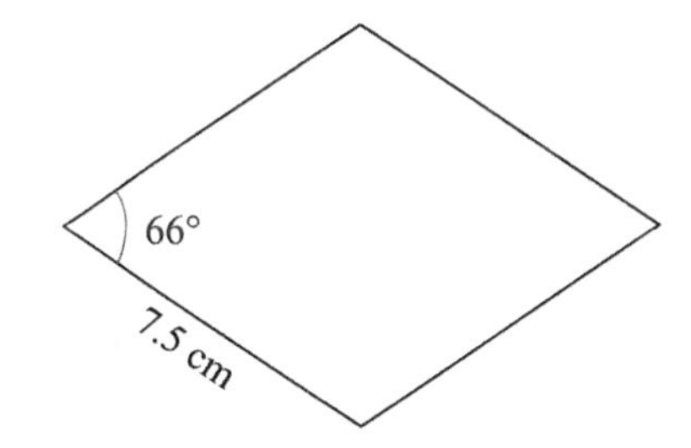

Puzzle corner. a. 1/4   b. 10   c. 6/15 or 2/5

Geometry Review, p. 141

1. The area of the triangle is exactly half of the area of the parallelogram.

2. a. 15 square units
   b. $20 - 2 - 1 - 4 - 3 = 10$ square units

3. a. $6.5\text{ m} \times 8.5\text{ m} + 6.5\text{ m} \times 4.5\text{ m} \div 2 = 55.25\text{ m}^2 + 14.625\text{ m}^2 = 69.875\text{ m}^2$

   b. The area with green beans is $3.5\text{ m} \times 3\text{ m} = 10.5\text{ m}^2$. It is 10.5 / 69.875 = 15% of the total garden area.

4. a. $13\text{ cm} \times 8.2\text{ cm} \div 2 = 53.3\text{ cm}^2$

   b. $130\text{ mm} \times 82\text{ mm} \div 2 = 5{,}330\text{ mm}^2$ (you can multiply the previous result by 100).

5. a. Check the student's work, as the nets can vary; for example:

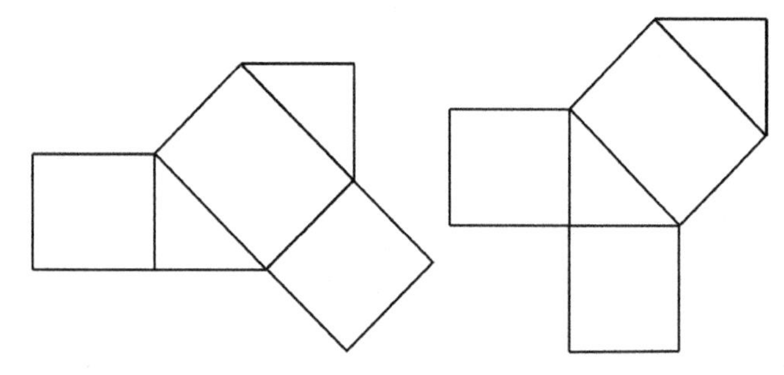

   Surface area: The two triangles are $2 \times 19\text{ in} \times 19\text{ in} \div 2 = 361$ sq. in.
   The rectangles: $27\text{ in} \times 21\text{ in} = 567$ sq. in.; $2 \times 19\text{ in} \times 21\text{ in} = 798$ sq. in.
   Total: $361\text{ in}^2 + 567\text{ in}^2 + 798\text{ in}^2 = 1{,}726\text{ in}^2$

   b.

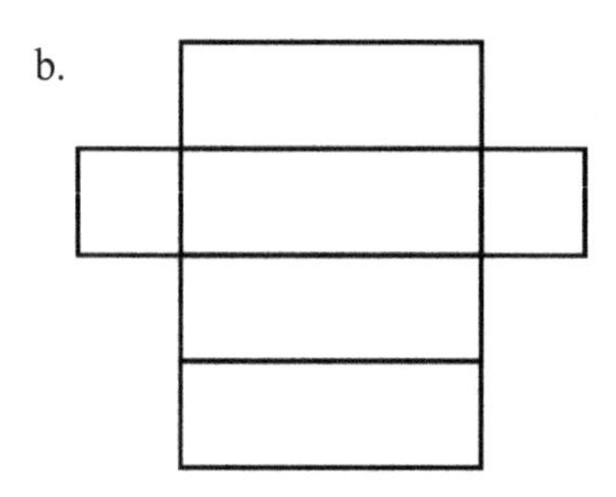

   Surface area: $4 \times 11.75\text{ in} \times 4\text{ in} + 2 \times 4\text{ in} \times 4\text{ in} = 188$ sq. in. + 32 sq. in. = 220 sq. in.

6. a. A triangular prism.
   b. A rectangular pyramid.
   c. A tetrahedron (triangular pyramid).

7. A triangular prism; Surface area: The triangles: $2 \times 3\text{ in} \times 2\ 5/8\text{ in} \div 2 = 3\text{ in} \times 21/8\text{ in} = 63/8$ sq. in. = 7 7/8 sq. in.
   The lateral sides: $3 \times 3\text{ in} \times 17\text{ in} = 153$ sq. in. Total: 160 7/8 sq. in.

8. The figure has 40 cubes, and the volume of each cube is $1/27\text{ m}^3$. The total volume of the figure is $40/27\text{ m}^3$ $= 1\ 13/27\text{ m}^3$. You can also calculate it by multiplying the three dimensions: $2/3\text{ m} \times 5/3\text{ m} \times 4/3\text{ m}$.

9. The volume of one story is $6\text{ m} \times 12.2\text{ m} \times 8.5\text{ m} \div 3 = 207.4\text{ m}^3$.

10. $50\text{ cm} \times 30\text{ cm} \times 40\text{ cm} \div 5 \times 4 = 48{,}000\text{ cm}^3 = 48{,}000\text{ ml} = 48\text{ L}$.

Chapter 10: Statistics

Understanding Distributions, p. 148

1. a. No. It could be changed to, for example: What color eyes do the teachers in my school have?
   b. Yes.
   c. Yes.
   d. Yes.
   e. No. It could be changed to, for example: What is the average wage in Ohio?
   f. No. It could be changed to, for example: On average, how many sunny days are there in London in August?
   g. No. It could be changed to, for example: How many pets do the children in my class have?

2. a. Bell-shaped, slightly asymmetrical or right-tailed (also called right-skewed).
   b. The peak is at 145 to 149 cm.
   c. 24

3.

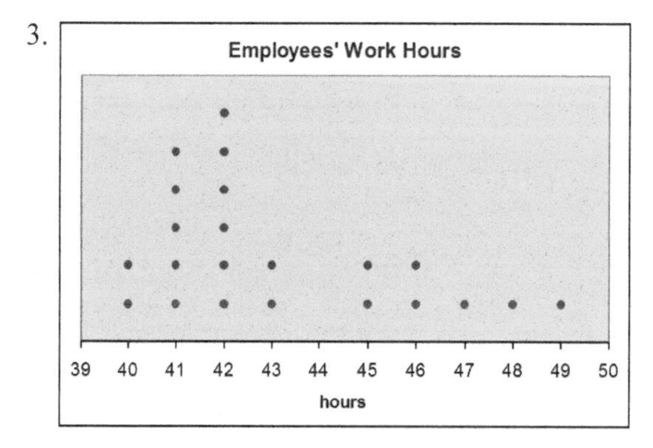

   a. The distribution is right-tailed (right-skewed).
   b. The peak is at 42 hours.
   c. 22 observations

4.

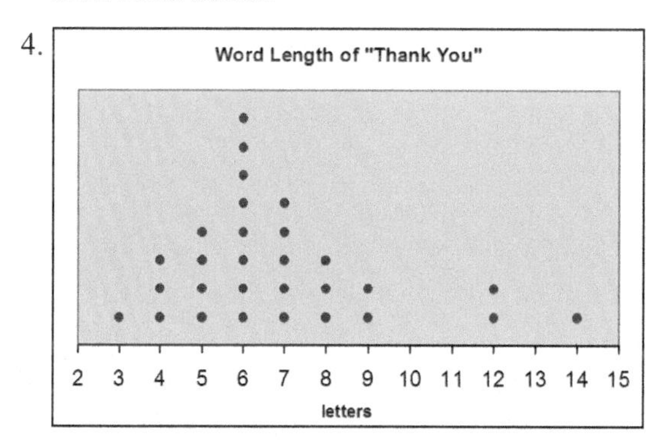

   a. Overall, this is bell-shaped. It is slightly skewed to the right or right-tailed.
   b. The peak is at 6 letters.
   c. 29

5.

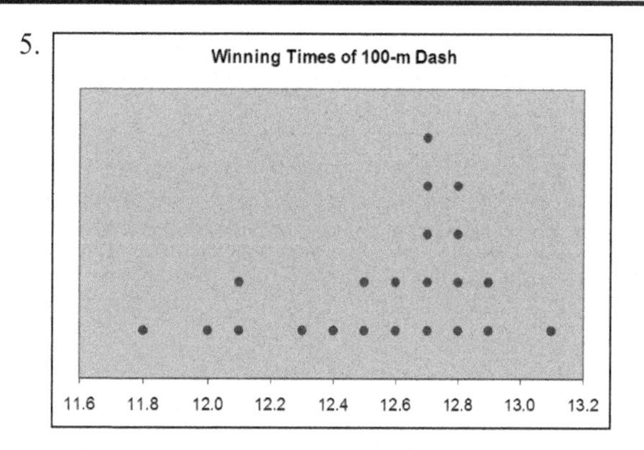

   a. The distribution is left-tailed (left-skewed).

   b. The peak is at 12.7 seconds.

   c. It has a medium spread. All of the athletes finished within about two seconds of time. However, the spread can appear wider if the interval for the horizontal axis in the graph is narrower (such as from 11.5 to 13.5 seconds). Also, whether the distribution appears to have a wide, medium, or narrow spread depends somewhat on the context. Maybe for the coach of these athletes the spread is wide, but for the general public the spread seems narrow.

   d. 22

6. a. About 55,000,000 people earned less than $12,500.

   b. About 45,000,000 people earned between $12,500 and $25,000.

   c. About 28% of the people earned less than $12,500. Since it asks for the approximate percentage, round the total number of people to 200,000,000. Then, take the fraction 55,000,000/200,000,000 = 55/200 and simplify it to 27.5/100, which equals 27.5%.

   d. About 23% of the people earned between $12,500 and $25,000. Similarly, we can look at the fraction 45,000,000/200,000,000 = 45/200 = 22.5/100.

Understanding Distributions, cont.

7. a.

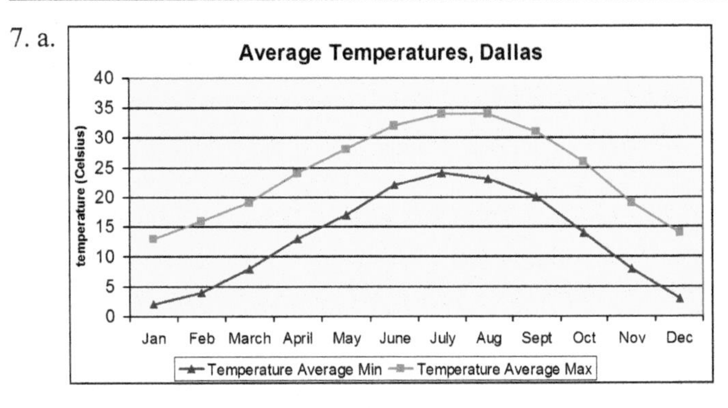

b. The temperature difference is 21 degrees.
c. The temperature difference is 22 degrees.
d. The difference between the maximum and minimum temperatures varies very little during the year (from 10 to 12 degrees). It is at its greatest in February and in October (12 degrees difference).

8. a. Yes. We are looking at the color of the candies, and that varies. There is no single answer to the question, "What color is the chocolate candy?"
b. Red: 10/55 = 0.181818.. ≈ 18%
Green: 7/55 = 0.127272727... ≈ 13%

Mean, Median, and Mode, p. 153

1. a. Median: 25. Mode: 25.
b. Median: 2. Mode: 2.
c. Median: 82.5. Mode: 80.
d. The median doesn't exist. There are two modes: crocs and sandals.

2. Joe's average time was 29.52 seconds.

3. a.

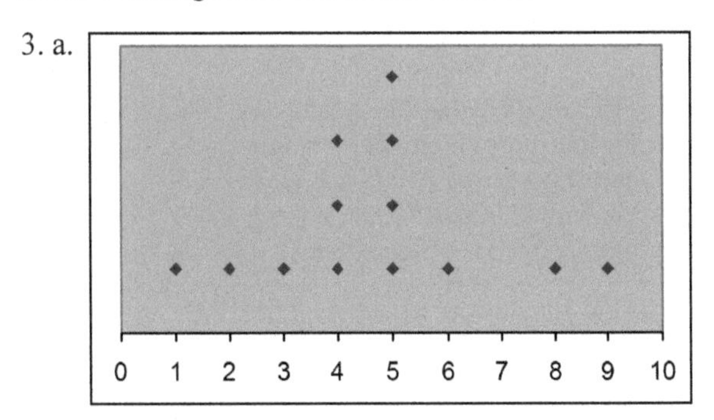

Mean: 4.69. Median: 5. Mode: 5.

b.

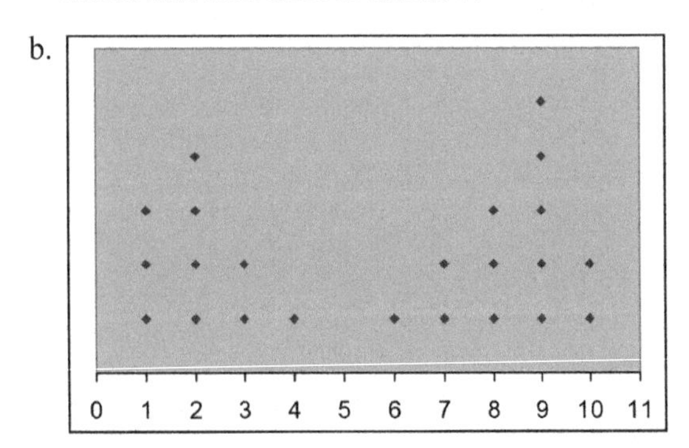

Mean: 5.65. Median: 7. Mode: 9.

This distribution is double-peaked.

4. a.

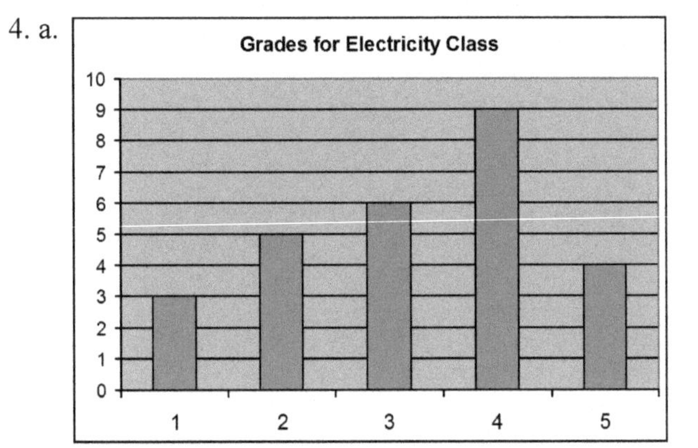

b. Mean 3.22. Median: 3. Mode: 4.
c. This is not an easy question. The mode describes the data well, but so does the mean. The mean is clearly better than the median, because it is closer to the peak of the distribution.

The mode does not tell us much about how the other students faired; it only tells us that 4 was the most common grade. Even a distribution where 14 students got 4 and 13 students got 1 would give a mode of 4, but in that case, reporting the mode as 4 would leave a false impression of most students doing well. That is why mode is rarely used. So, mean is probably the best choice.

The best practice is to always show a distribution visually, not only to report a measure of center.

Using Mean, Median, and Mode, p. 155

1. a. Mean: 7. Median: 5. Mode: 5.
   b. Clearly, either the median or the mode, but *not* the mean! The mean is off from the central peak of the distribution.

2. ____ mean ____ median _X_ mode
   Only the mode is possible because the original data consists of just the words "yes" and "no." (For example, the data could have been: yes, yes, yes, no, no, yes, no, yes, yes, no, no ...)

3. a. mean is $11; median is $9
   The majority of the data is in the first bar, $1-$10, and the distribution is totally skewed (not bell-shaped at all), so the mean will not describe it as well as the median does. The mean is "thrown off" by the few large numbers in the distribution whereas the median is not. Therefore, the mean must be the larger of the two numbers given ($11), and the median is $9.
   b. The median is better for describing this data, because its value, $9, falls into where the peak of the data is (the first bar on the graph).
   c. About 50% of the teens spent less than $10 on a gift.

4. a. What is being measured or studied?
   How many hours certain people sleep.

   How is it measured?
   By asking the people to tell how many hours they slept.

   Which are possible?
   _X_ mean _X_ median _X_ mode

   The mode is 9 hours. The median is 9 hours

4. b. What is being measured or studied?
   Eye colors of students

   How is it measured?
   Perhaps by observing their eyes closely, or perhaps by asking them what color their eyes are.

   Which are possible?
   ____ mean ____ median _X_ mode

   The mode is blue. There is no median.

5. a.

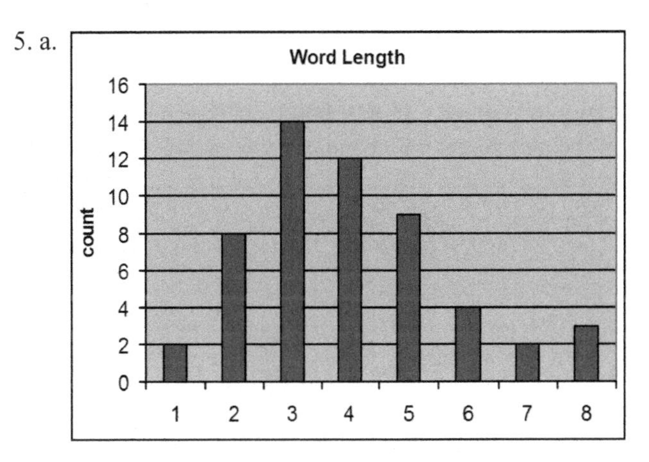

This distribution is bell-shaped and right-tailed.

There are 54 observations. The mean is 3.98. The median is 4. The mode is 3. Any one of them could be used, though in this case, the mode describes the peak of the distribution best.

5. b.

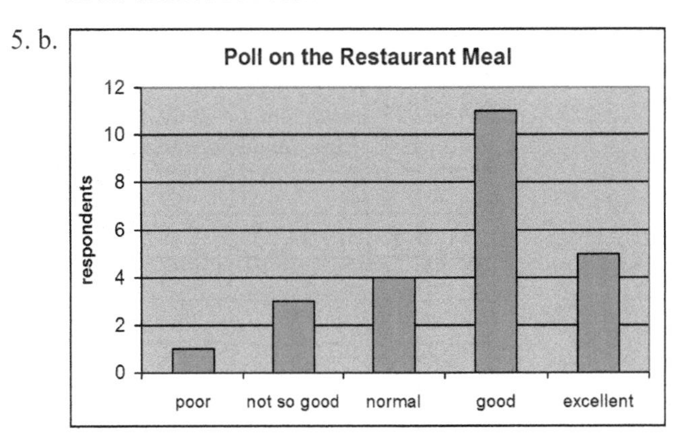

The distribution is left-tailed (left-skewed).

There were 24 observations (respondents). The mode is "good." (Note, the mean and the median are not possible.)

Puzzle corner:
For a quick answer, look at each number's difference from 100. The numbers greater than 100 have a total surplus of: $2 + 5 + 7 + 1 + 4 = 19$. The numbers less than 100 lack: $6 + 1 + 4 = 11$. That gives a net difference of $19 - 11 = 8$. To get an average, that net difference needs to be distributed among the 8 numbers, so each gets a surplus of 1, and the mean is thus 101 grams.

Measures of Variation, p. 158

1. a. First quartile: 6.5. Median: 7. Third quartile: 8. Interquartile range: 1.5.

   b. First quartile: 4.5. Median: 5.5. Third quartile: 7. Interquartile range: 2.5.

   c. The first group did better in general (look at the graphs or at the medians).
   The quiz scores varied more in the second group (look at the graphs or the interquartile ranges).

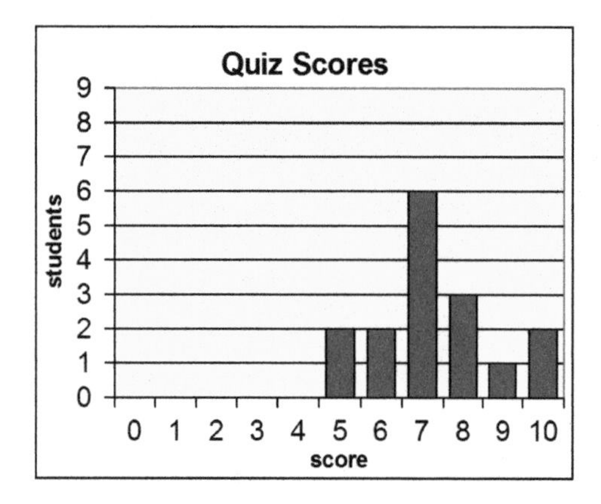

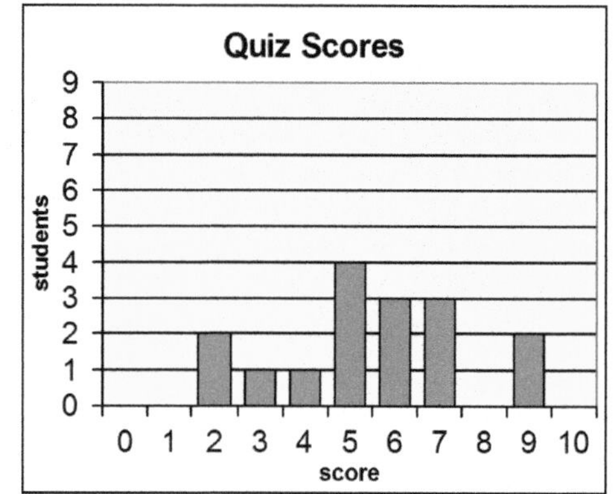

2. a. Range: 18 days. 1st quartile: 11 days. Median: 13 days. 3rd quartile: 15 days. Interquartile range: 4 days.
   b. Range: 11 cm. 1st quartile: 139.5 cm. Median: 140 cm. 3rd quartile: 143 cm. Interquartile range: 3.5 cm.

3. a.

Art Club – members' ages			
	age	difference from mean	absolute difference
	7	-5	5
	9	-3	3
	9	-3	3
	10	-2	2
	12	0	0
	13	1	1
	13	1	1
	13	1	1
	14	2	2
	14	2	2
	15	3	3
	15	3	3
mean	12	***m.a.d.***	2.167

b.

Prices of MP3 players			
	price	difference from mean	absolute difference
	29	-6	6
	30	-5	5
	33	-2	2
	34	-1	1
	34	-1	1
	35	0	0
	35	0	0
	35	0	0
	36	1	1
	36	1	1
	37	2	2
	39	4	4
	42	7	7
mean	35	***m.a.d.***	2.308

4. a.

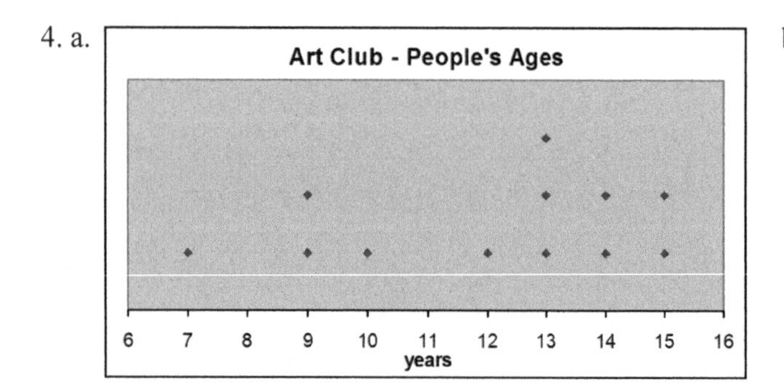

b.

Prices of different MP3 players

25 30 35 40 45

dollars

Measures of Variation, cont.

5. a. The distribution is left-tailed. The median would describe the center of this data better than the mean. Interquartile range should be chosen as the measure of variation.

   b. The distribution is bell-shaped. The mean works well for describing this data (though the median does, too). If you choose the mean, then use mean absolute deviation as the measure of variation. If you choose the median, then use interquartile range.

Making Histograms, p. 163

1.

Height (cm)	Frequency
154 - 158	5
159 - 163	7
164 - 168	7
169 - 173	2
174 - 178	3

Height of Swimmers

frequency

0 1 2 3 4 5 6 7 8

154..158 159..163 164..168 169..173 174..178

cm

2.

Score	Frequency
60 - 65	2
66 - 71	2
72 - 77	12
78 - 83	10
84 - 89	3
90 - 95	3

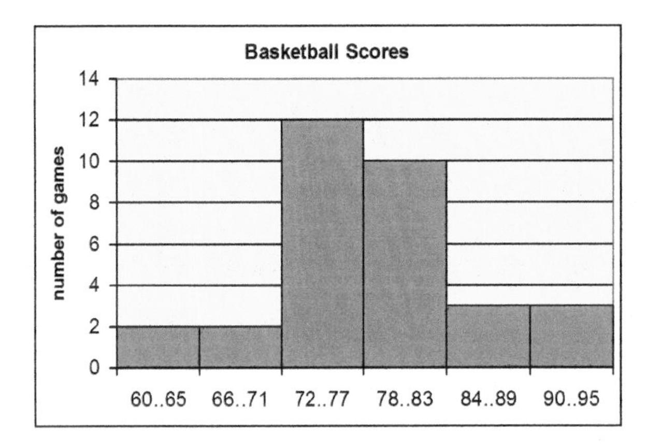

3. a. median 77.5 interquartile range 82.5 − 74 = 8.5
   b. It is somewhat scattered.

4.

Life expectancy (years)	Frequency
65.0 - 67.9	2
68.0 - 70.9	2
71.0 - 73.9	5
74.0 - 76.9	4
77.0 - 80.9	1

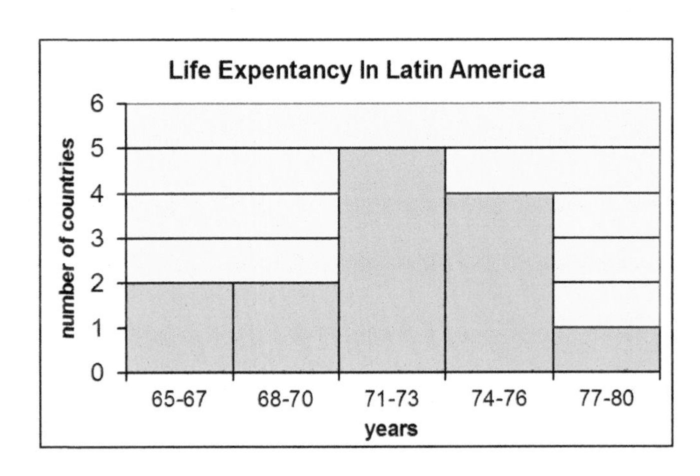

Making Histograms, cont.

5. The histograms may vary greatly depending on the choice of bins. Check the student's work. For example, if the student chooses bins 11.0-11.9, 12.0-12.9, 13.0-13.9, and 14.0-14.9, the frequencies will be 1, 20, 1, and 0. Or if the student chooses bins 11.5-11.9, 12.0-12.4, 12.5-12.9, and 13.0-13.4, the frequencies will be 1, 5, 15, and 1. The whole idea is to notice how much the choice of bins affects the look of the histogram when we do not have very many data items, and thus, a dot plot or a stem-and-leaf plot that shows all the individual data items might work better.

   An example answer is below.

5. a.

Winning times (seconds)	Frequency
11.6-11.9	1
12.0-12.3	3
12.4-12.7	10
12.8-13.1	7

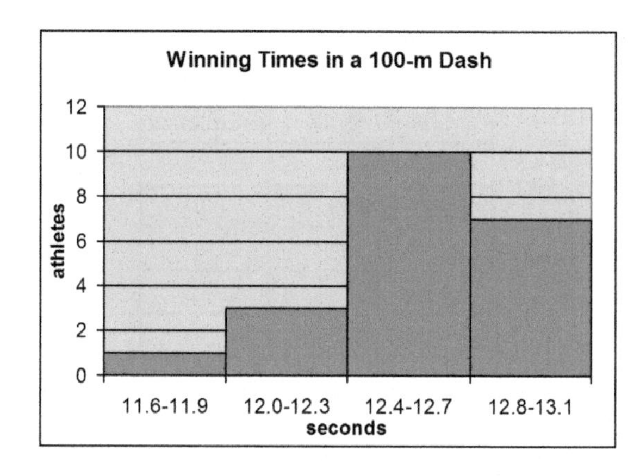

b.

Winning times (seconds)	Frequency
11.7-11.9	1
12.0-12.2	3
12.3-12.5	4
12.6-12.8	11
12.9-13.1	3

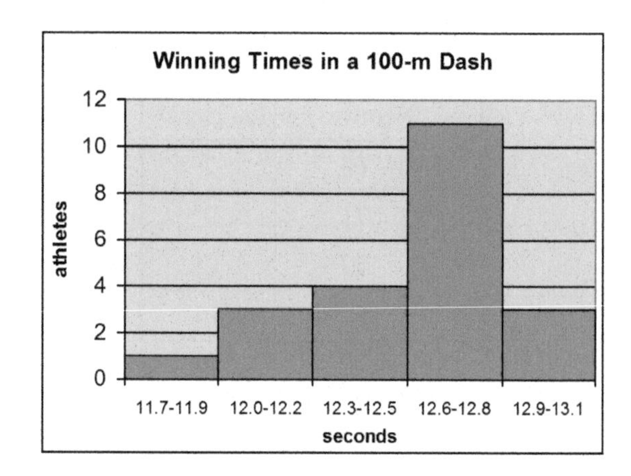

Boxplots, p. 166

1. a. Minimum: 8 points
   First quartile: 12 points
   Median: 15 points
   Third quartile: 17 points
   Maximum: 19 points

   b. The interquartile range is 5 points.

2. a. Minimum: 2 days
   First quartile: 5.5 days
   Median: 7 days
   Third quartile: 8.5 days
   Maximum: 16 days

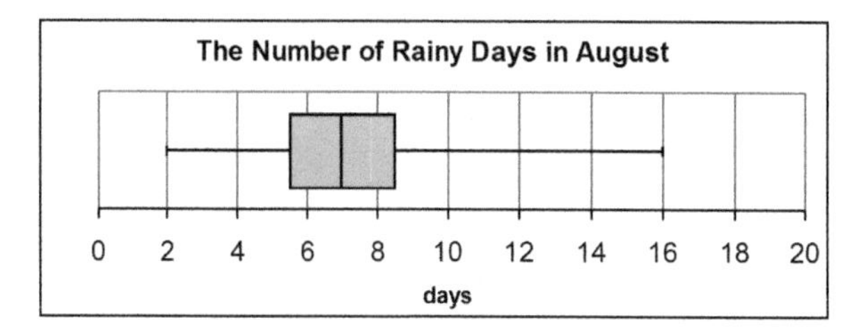

2. b. Minimum: \$23
   First quartile: \$29
   Median: \$30
   Third quartile: \$31
   Maximum: \$38

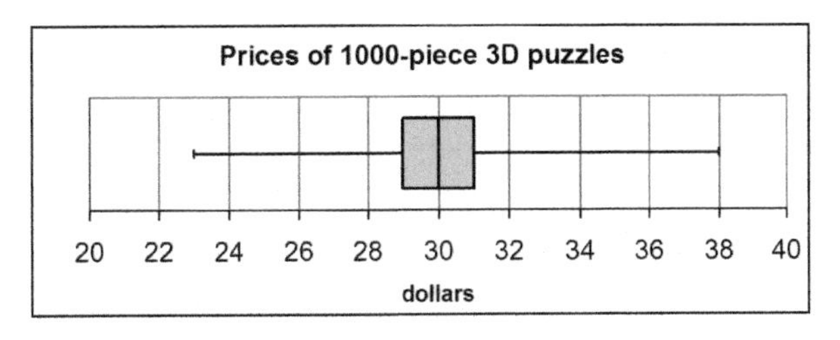

2. c. Minimum: 46 points
   First quartile: 63 points
   Median: 70 points
   Third quartile: 74 points
   Maximum: 85 points

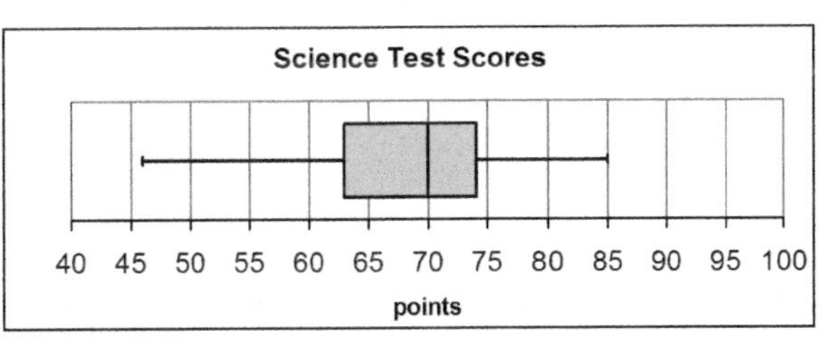

3. a. Group 2 did the best over all. The median for that group is the largest of the three.
   b. Group 3 varied the most, as the range for that group is the largest of the three.
   Group 1 varied the least. Again, the range tells us this: the range for this group is the smallest of the three.
   c. Group 1 has the smallest interquartile range (about 9, versus 13 and 11).

4. The best choices are a histogram or a boxplot. This data really needs to be grouped before making a plot, so a dot plot does not work well.

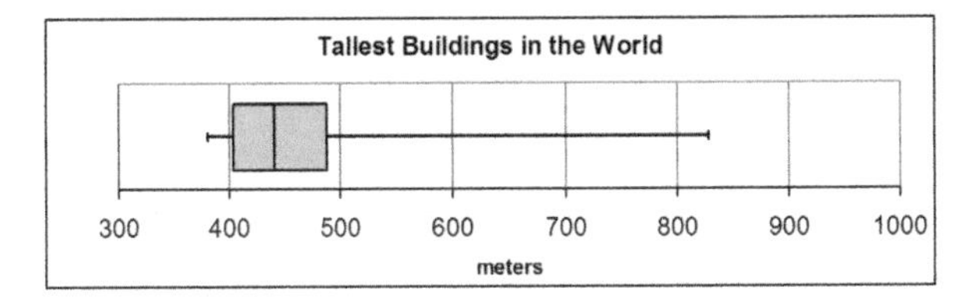

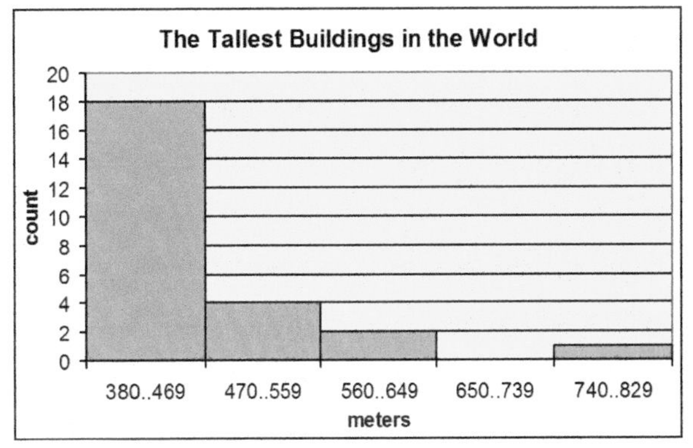

One could also make a bar graph with an individual bar for each building, and the building's name underneath, something like this:

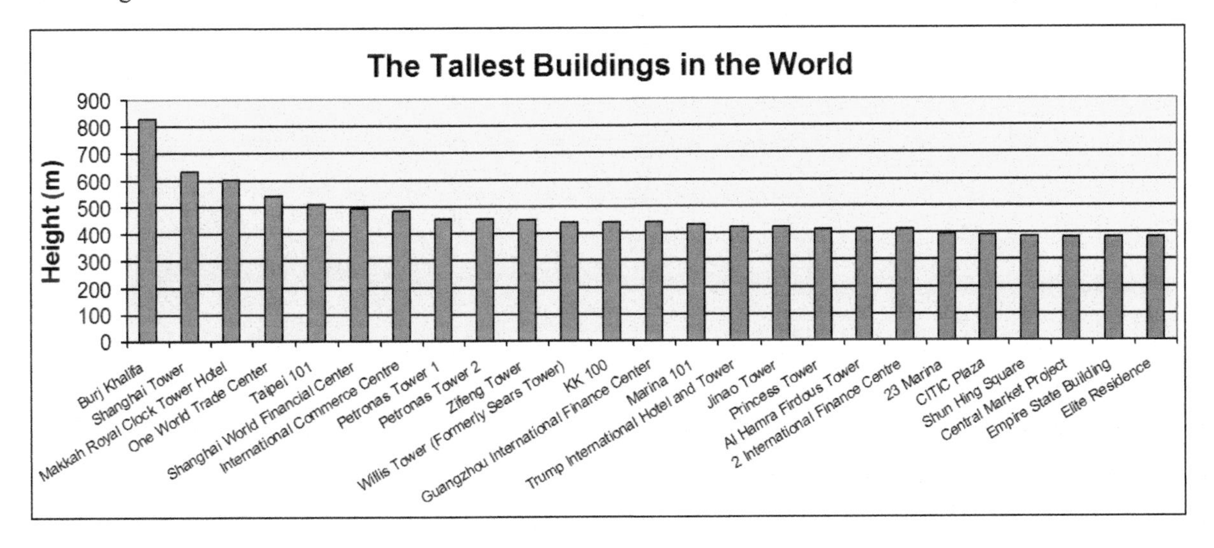

However, that was not the intention of the exercise.

Stem-and-Leaf Plots, p. 170

1. a.

Stem	Leaf
1	4 5 9
2	0 0 0 1 4 5
3	4 4 5 5 7
4	2 5
5	5

   b. The median is 25.

2. a.

Stem	Leaf
70	0 9
71	9
72	5 8
73	
74	0 5
75	0 5 7
76	
77	0
78	6
79	

   b. The median monthly rent is $742.50.

   c. The first quartile is (725 + 719) / 2 = 722.
   The third quartile is (755 + 757) / 2 = 756.
   The interquartile range is 756 − 722 = 34.

   d. The data is spread out a lot.

3. a.

Stem	Leaf
4	9
5	2 9
6	3 3 7 9
7	1 3 7 9 9
8	2 4
9	0 6
10	5 6

   b. The median is 75.
   The 1st quartile is 63. The 3rd quartile is 84.
   The interquartile range is 21.

4. a. The median of the scores is 9.5.
   b. The mean of the scores is 9.43.
   c. the median

Stem	Leaf
9	1 2
9	3 3
9	5 5 5 6 6
9	7

5. a.

Stem	Leaf
22	7
23	1 2 5 8
24	8
25	0 4 5 9
26	0

   b. It is double-peaked.

   c. None of them. There is no mode. The median is 248, and the mean is 244.45, so neither of them falls at either peak.

6. Check the student's answers.

Mixed Review, p. 173

1. a. 3 b. 16

2. a. 12 b. 24

3.

a. The GCF of 72 and 12 is 12.	b. The GCF of 42 and 66 is 6.
12 + 72 = 12 (1 + 6)	42 + 66 = 6 (7 + 11)

4. a.

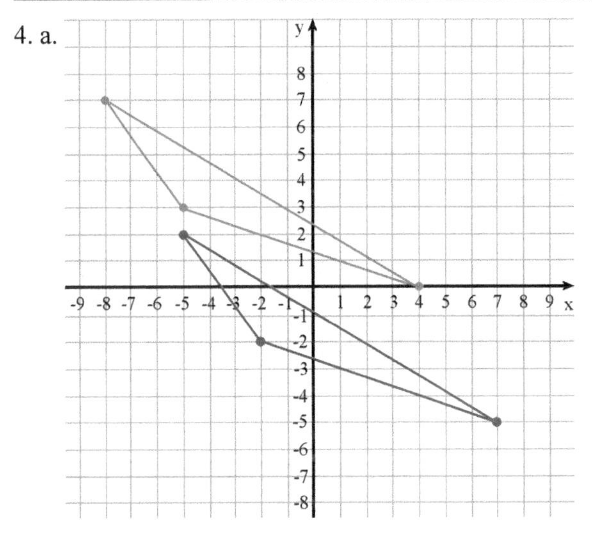

b. (−5, 2), (−2, −2) , and (7, −5)

5. a. $19s = 304$. $s = 304 \div 19 = 16$.
   The other side is 16 m.

   b. 5w = 6.7. w = 6.7 ÷ 5 = 1.34.
   Each book weighed 1.34 kg.

6. a. $7z^3$
   b. $240ab$
   c. $4x + 2$
   d. $3t + 6$

7. a. (30 + 33 + 55)/66 = 1 52/66 = 1 26/33

   b. 3 55/60 − 30/60 + 15/60 = 3 40/60 = 3 2/3

8. (2/3) × (9/10) = 18/30 = 3/5 of the whole pizza

9. The piglet gains 12 × 7 1/3 oz = 88 oz = 5 lb 8 oz.
   Its final weight is 3 lb 4 oz + 5 lb 8 oz = 8 lb 12 oz.

10. 5 3/4 in ÷ 4 = 23/4 in ÷ 4 = 23/16 in = 1 7/16 in.

11. a. 9
    b. 3
    c. 0
    d. 28
    e. 7

12. a. $^-12 + 7 = {^-5}$
    b. $2 - 8 = {^-6}$

13. Check the student's drawing. The product of the lengths of the perpendicular sides should be 16. For example, the two perpendicular sides could measure 4 in and 4 in, or 2 in and 8 in, or 5 1/3 in and 3 in, *etc.*

14. Sail: 3 × 6 ÷ 2 = 9 square units.
    Boat: 1 + 16 + 1 = 18 square units.
    Total: 27 square units.

15. 19 in × 11 in − 10 in × 5.5 in ÷ 2
    = 209 in^2 − 27.5 in^2 = 181.5 in^2.

16. a. The three warmest months are June, July, and August. The three coldest months are January, February, and December.
    b. 27°C − 2°C = 25°C
    c. In August, 9 degrees. In January, also 9 degrees.

Puzzle corner.
a. Jerry "flipped" the first fraction, or used the reciprocal of the first fraction, and then multiplied.
b. Emily ignored the whole-number part of the mixed numbers.

1. a. Yes.
   b. No. Change it to (for example): "How many pages are in the math books in the library?" Or: "How many pages are in my problem-solving books?"

2. a. About 97% of the population of Norway uses the Internet.
   b. About 84% of the population of the United Kingdom uses the Internet.
   c. So there are about 15.0 million Internet users in the Netherlands.
   (Multiply 0.89 × 16,847,000 and round to the nearest tenth of a million.)
   d. So there are about 4.6 million Internet users in Finland.
   (Multiply 0.88 × 5,260,000 and round to the nearest tenth of a million.)

3. Mean: 7.6. Median: 7.5. Mode: 7. (The original data is 4, 6, 6, 7, 7, 7, 7, 7, 7, 8, 8, 8, 8, 9, 9, 9, 10, 10.)
   b. It is more or less bell-shaped, though slightly left-tailed (left-skewed).

4. a. Minimum: 2.
   First quartile: 5.5.
   Median: 7.
   Third quartile: 8.
   Maximum: 12.
   Interquartile range: 2.5.

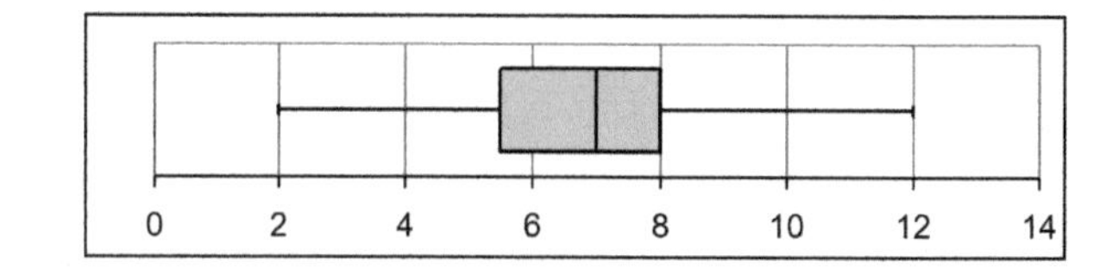

   b. Answers will vary. It could be the ages of some group of children, such as children playing in the park.

5. a.

Stem	Leaf
6	4 6
7	5 7 8
8	2 4 4 5
9	0
10	
11	2

   b. Median: 82.
   c. Range: 112 − 64 = 48.
   d. The number 112 is the outlier.
   e. Without the outlier, it is J-shaped.

6. a. about 1/2 of the people
   b. It is J-shaped.
   c. The median is, because it falls within the highest bar, whereas the mean does not.

7. a.

California Cities Average Annual Rainfall

0 5 10 15 20 25 30 35 40 45 50 55 60 65 70 75
inches

   b. The data is spread out a lot.
   c. The distribution is more or less bell-shaped and it has an outlier.
   d. Median: 14 inches. Note that mean is not a good measure of center to use here because of the outlier. The outlier throws the mean off.

   e.

Rainfall (in)	Frequency
5 - 20	11
21 - 36	1
37 - 52	2
53 - 69	1

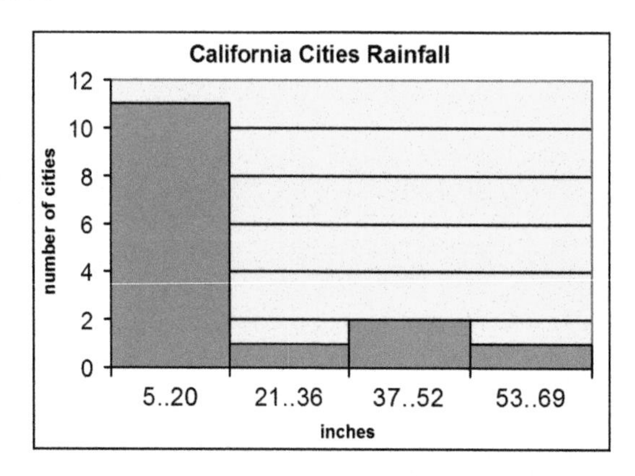

Test Answer Keys

Math Mammoth Grade 6 Tests Answer Key

Chapter 1 Test

Grading

My suggestion for grading is as follows. The total is 23 points. You can give partial points for partial solutions. Divide the student's score by 23 and multiply by 100 to get a percent score. For example, if the student scores 17, divide 17 ÷ 23 with a calculator to get 0.7391. The percent score is 73.9% or 74%.

Question #	Max. points	Student score
1	2 points	
2	2 points	
3	2 points	
4	4 points	
5	4 points	

Question #	Max. points	Student score
6	2 points	
7	2 points	
8	2 points	
9	3 points	
TOTAL	**23 points**	/ 23

1. 10,540 R28

2. 1.909

3. \$937.50 ÷ 75 = \$12.50. One flashlight costs \$12.50.

4. It will take you 43 sec/page × 234 pages = 10,062 seconds = 167 minutes and 42 seconds (almost 3 hours).

5. a. $3^3 = 27$
   b. $1^{10} = 1$
   c. $50^2 = 2{,}500$
   d. $10^5 = 100{,}000$

6. The length of each side is 56 cm ÷ 4 = 14 cm. So the area is 14 cm × 14 cm = 196 cm^2.

7. a. 504,300,000
   b. 1,600,020,100

8. a. $5 \times 10^5 + 6 \times 10^4$
   b. $9 \times 10^6 + 1 \times 10^5 + 8 \times 10^3$

9. a. 3,000,000
   b. 480,000,000
   c. 20,000,000

Chapter 2 Test

Grading

My suggestion for grading is as follows. The total is 30 points. You can give partial points for partial solutions. Divide the student score by 30 and multiply by 100 to get a percent score. For example, if the student scores 25, divide 25 ÷ 30 with a calculator, getting 0.833333.... The percent score is 83%.

Question #	Max. points	Student score
1	3 points	
2	3 points	
3	4 points	
4	2 points	
5	4 points	
6	2 points	

Question #	Max. points	Student score
7	3 points	
8	2 points	
9	2 points	
10	2 points	
11	3 points	
TOTAL	30 points	/30

1. a. $x^2/7$ b. $(5 - y)^3$ c. $3(2s - 5)$

2. a. 20 b. 60 c. 32

3. a. 20 b. 35 c. 7 d. 39

4. $p + 3t$

5. a. a^4 b. $4a$ c. $10x^2$ d. $6d + 7$

6. a. $5(x + 6) = 5x + 5 \cdot 6 = 5x + 30$
   b. $2(9 + 5y) = 2 \cdot 9 + 10y = 18 + 10y$

7. a. $x = 144 \div 6 = 24$
   b. $y = 134 - 78 = 56$
   c. $x = 3 \cdot 16 = 48$

8. Let one side be denoted by s. The equation is $4s = 164$. Solution: $s = 41$

9.

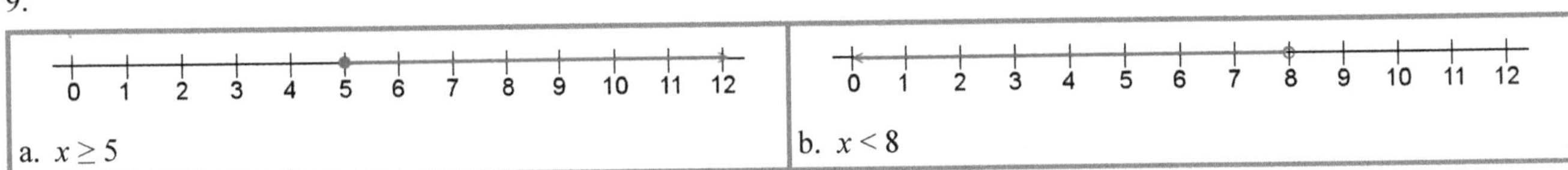

10. 15, 16

11. $y = 9 - x$

x	0	1	2	3	4	5	6	7	8	9
y	9	8	7	6	5	4	3	2	1	0

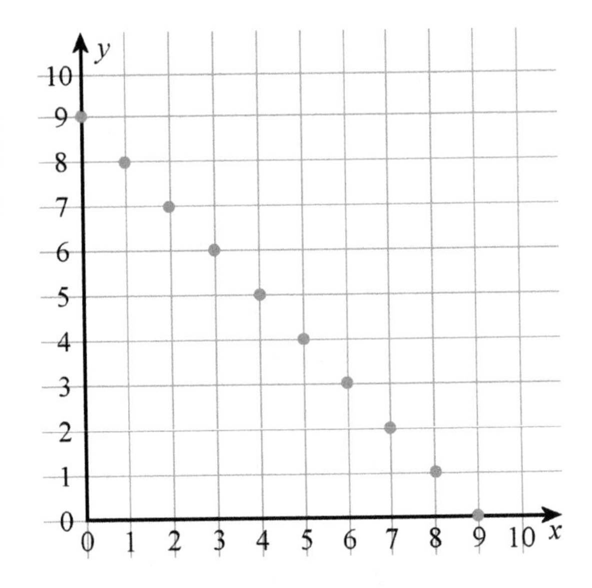

Chapter 3 Test

Grading

My suggestion for grading is as follows. The total is 66 points. You can give partial points for partial solutions. Divide the student score by 66 points and multiply by 100 percent to get a percent score. For example, if the student scores 51, divide 51 ÷ 66 with a calculator, getting 0.77272727. The percent score is 77%.

Question #	Max. points	Student score
1	6 points	
2	3 points	
3	3 points	
4	4 points	
5	6 points	
6	6 points	
7	4 points	

Question #	Max. points	Student score
8	6 points	
9	3 points	
10	4 points	
11	9 points	
12	2 points	
13	2 points	
14	8 points	
TOTAL	66 points	/ 66

1. a. 0.005 b. 0.00382 c. 1.003658
   d. 0.0094 e. 0.65 f. 8.08

2. a. 20,045 / 10,000 b. 912 / 1,000,000 c. 749,038 / 100,000

3. The student may give the answer either as a decimal or as a fraction — both are correct.
   a. 0.2 + 0.005 = 0.205 OR 205/1000
   b. 0.07 + 0.03 = 0.10 = 0.1 OR 1/10
   c. 2.022 + 0.033 = 2.055 or 2 55/1000

4. a. 2.31 × 0.04 = 0.0924 b. 3.38758 ÷ 7 + 0.045 = 0.52894

5.

Round to …	0.0882717	0.489932	1.299959
… the nearest thousandth	0.088	0.490	1.300
… the nearest hundred-thousandth	0.08827	0.48993	1.29996

6. a. 0.24 ÷ 3 = 0.08 b. 5.4 ÷ 0.6 = 9 c. 0.081 ÷ 0.009 = 9
   d. 2 × 0.05 = 0.1 e. 8 × 0.009 = 0.072 f. 11 × 0.0005 = 0.0055

7. The area of the parcel is 50.5 m × 27.6 m = 1,393.8 m^2, so each fourth has an area of 1,393.8 m^2 ÷ 4 = 348.45 m^2.

8. a. 20 b. 0.47 c. 1097000 d. 0.006 e. 0.001245 f. 0.00324

9. a. 0.04 ÷ 4 = 0.01 b. 0.04 ÷ 0.04 = 1 c. 0.04 ÷ 10 = 0.004

10. a. 56 mm = 0.056 m b. 9 km = 9,000 m c. 9 cg = 0.09 g d. 16 dl = 1.6 L

11. a. 2.7 km = 2,700 m = 270,000 cm = 2,700,000 mm
    b. 5,600 ml = 560 cl = 56 dl = 5.6 L
    c. 0.6 g = 6 dg = 60 cg = 600 mg

12. Since 7 pounds 6 ounces = 7 6/16 lbs = 7.375 lbs, 7.4 pounds is heavier than 7 pounds 6 ounces by 7.4 lbs − 7.375 lbs = 0.025 lbs (= 1/40 lb = 4/10 ounce).

13. The pint-sized bottle is $7/16 oz = 43.75¢ per ounce, and the 24-ounce size is $12/24 oz = 50¢ per ounce. The pint-sized bottle of honey is the better deal. You can also solve this by figuring out the price for 48 ounces: Three pint-sized bottles (48 oz in total) cost $21, whereas two of the 24-ounce bottles cost $24. The pint-sized bottles are the better deal.

14. a. 5.36 ÷ 0.2 = 26.8 b. 1.6 ÷ 0.05 = 32
    c. 22.9 ÷ 7 = 3.271 d. $\frac{8}{9}$ = 0.889

Chapter 4 Test

Grading

My suggestion for grading is as follows. The total is 35 points. You can give partial points for partial solutions. Divide the student score by 35 points and multiply by 100 percent to get a percent score. For example, if the student scores 25, divide 25 ÷ 35 with a calculator to get 0.71428. The percent score is 71%.

Question #	Max. points	Student score
1	4 points	
2a	1 point	
2b	2 points	
3	5 points	
4a	1 point	
4b	2 points	
5	2 points	
6a	1 point	
6b	2 points	

Question #	Max. points	Student score
7	3 points	
8a	1 point	
8b	2 points	
8c	2 points	
9	3 points	
10	4 points	
TOTAL	35 points	/ 35

1. a. 3/5 = 18/30 b. 2:3 = 18: 27 c. 10 to 45 = 2 to 9 d. 12:30 = 2 to 5

2. a. 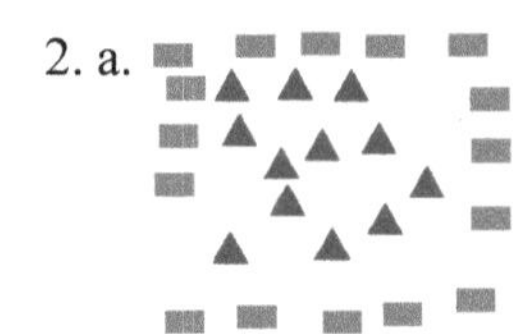

b. 1 1/3 rectangles for **1** triangle (the ratio of rectangles to triangles is 16:12 = 4:3 = 4/3 : 1)
3/4 triangles for **1** rectangle (the ratio of triangles to rectangles is 12:16 = 3:4 = 3/4 : 1)

3. a. $\dfrac{4\text{ L}}{10\text{ m}^2} = \dfrac{2\text{ L}}{5\text{ m}^2} = \dfrac{10\text{ L}}{25\text{ m}^2}$ b. $\dfrac{\$9}{6\text{ min}} = \dfrac{\$3}{2\text{ min}} = \dfrac{\$15}{10\text{ min}} = \dfrac{\$90}{1\text{ hour}}$

4. a. The unit rate is 0.1 m per 1 minute.
b. 17 min · 0.1 m/min = 1.7 meters.

5. a. $\dfrac{14\text{ downloads}}{\$2.10} = \dfrac{2\text{ downloads}}{\$0.30} = \dfrac{1\text{ download}}{\$0.15} = \dfrac{3\text{ downloads}}{\$0.45}$, so three song downloads would cost 45¢.

6. The ratio is 8:5 and the shorter side length of the rectangle is 15 cm, so each "part" in the ratio corresponds to 3 cm.
a. The rectangle's length is 8 · 3 cm = 24 cm.
b. Its area is 15 cm · 24 cm = 360 cm^2.

7. The 1 + 7 = 8 parts make a total of 4 L, so each part is 4 L ÷ 8 = ½ L. You need ½ L of juice concentrate and 7 × ½ L = 3 ½ L of water.

8. a. 35 gal per 7 mi = 5 gal per 1 mi.
b. The plane can fly 100 miles with 500 gallons of fuel.
c. The plane will need 750 gallons of fuel to travel 150 miles.

9. There are 3 + 5 = 8 parts, so each part consists of 1,200 ÷ 8 = 150 inserts. Anita folded 3 · 150 = 450 inserts and Michael folded 5 · 150 = 750 inserts.

10.

a. $60\text{ cm} = 60\text{ cm} \cdot \dfrac{1\text{ in}}{2.54\text{ cm}} = \dfrac{60\text{ in}}{2.54} \approx 23.6\text{ in}$
b. $4.5\text{ ft} = 4.5\text{ ft} \cdot \dfrac{30.48\text{ cm}}{1\text{ ft}} = 4.5 \cdot 30.48\text{ cm} = 137.16\text{ cm} \approx 137\text{ cm}$

Chapter 5 Test

Grading

My suggestion for grading is as follows. The total is 36 points. You can give partial points for partial solutions.
Divide the student score by 36 to get a percent score. For example, if the student scores 24, divide 24 ÷ 36 with a calculator, getting 0.6666666... The percent score is 67%.

Question #	Max. points	Student score
1	6 points	
2	9 points	
3	2 points	
4	2 points	
5	2 points	
6	2 points	
7	2 points	

Question #	Max. points	Student score
8	2 points	
9	2 points	
10	3 points	
11	2 points	
12	2 points	
TOTAL	36 points	/ 36

1.

a. 45% = $\frac{45}{100}$ = 0.45	b. 179% = $\frac{179}{100}$ = 1.79	c. 2% = $\frac{2}{100}$ = 0.02

2.

percentage / number	5,200	80	9
1% of the number	52	0.8	0.09
3% of the number	156	2.4	0.27
70% of the number	3,640	56	6.3

3. 57.1%

4. $8.40. First find 30% of $12. It is $3.60. Then subtract that from $12.00.

5. $8.40. First find 20% of $7. Since 10% of $7 is $0.70, then 20% of $7 is $1.40 Add that to the original price of $7.00.

6. Seventy-two T-shirts are not white. You can first calculate 10% of 120, which is 12. Then, 60% of 120 is six times as much, or 72.

7. Twenty percent of the caps are red. There are 2 red caps and 10 in total. Two caps is 2/10 or 20% of the caps.

8. Sixty-seven percent are boys. There are 16 boys, so in total, 16/24 = 2/3 = 67% of the members are boys.

9. 144 cm/160 cm = 9/10 = 90%. Annie's height is 90% of Jessie's height.

10. The $35 jeans discounted by 10% are cheaper. They are $0.50 cheaper.
To find the price of $35 jeans discounted by 10%: First find 10% of $35. It is $3.50. Then subtract $35 − $3.50 = $31.50.
To find the price of $40 jeans discounted by 20%: First find 20% of $40. It is $8. Then subtract $40 − $8 = $32.

11. His salary is $2,000. If $400 is 20%, then $200 is 10%, and $2,000 is 100% or Andrew's total salary.

12. The total population is 14,000. If 15% is 2,100, divide those by 3 to get that 5% is 700. Then, multiply that 700 by 20 to get 100% or the total population.

Chapter 6 Test

Grading

My suggestion for grading is as follows. The total is 23 points. You can give partial points for partial solutions. Divide the student score by 23 to get a percent score. For example, if the student scores 17, divide 17 ÷ 23 with a calculator, getting 0.739130... The percent score is 74%.

Question #	Max. points	Student score
1	6 points	
2	2 points	
3	2 points	
4	2 points	
5	2 points	

Question #	Max. points	Student score
6	1 point	
7	2 points	
8	4 points	
9	2 points	
TOTAL	23 points	/ 23

1. a. $2 \cdot 2 \cdot 2 \cdot 7$
   b. $2 \cdot 3 \cdot 3 \cdot 5$
   c. 101 is a prime number

2. a. 24
   b. 12

3. a. 2
   b. 7

4. Answers will vary. For example 30, 60, 90, and 120.

5. $40 \times 2 = 80$

6. 1 is a factor of all numbers.

7. They are prime numbers so their greatest common factor is 1.

8.

a. The GCF of 24 and 30 is 6. $24 + 30 = 6 \cdot 4 + 6 \cdot 5 = 6(4 + 5)$
b. The GCF of 22 and 121 is 11. $22 + 121 = 11(2 + 11)$

9.

a. $\frac{124}{72} = \frac{31}{18} = 1\frac{13}{18}$	b. $\frac{65}{105} = \frac{13}{21}$

Chapter 7 Test

Grading

My suggestion for grading is as follows. The total is 30 points. You can give partial points for partial solutions.
Divide the student score by 30 to get a percent score. For example, if the student scores 25, divide 25 ÷ 30 with a calculator, getting 0.833333... The percent score is 83%.

Question #	Max. points	Student score
1	8 points	
2	2 points	
3	2 points	
4	2 points	
5	2 points	

Question #	Max. points	Student score
6	2 points	
7	4 points	
8	3 points	
9	2 points	
10	3 points	
TOTAL	30 points	/30

1.

a. $\frac{5}{12} + \frac{6}{12} + \frac{10}{12} = \frac{21}{12} = \frac{7}{4} = 1\frac{3}{4}$	b. $\frac{35}{63} - \frac{18}{63} = \frac{17}{63}$
c. $2\frac{18}{60} + 2\frac{55}{60} = 4\frac{73}{60} = 5\frac{13}{60}$	d. $7\frac{3}{15} - 5\frac{7}{15} = 1\frac{11}{15}$

2. 1/4 × 3/4 = 3/16 of the original pizza left.
   Since Joe ate 3/4 of what had been left, Joe didn't eat 1/4 of what had been left. So, what remains is 1/4 of the 3/4 of the pizza. Or, you can calculate that the part Joe ate was 3/4 × 3/4 = 9/16 of the original pizza, and the family had eaten 1/4 = 4/16 of the original pizza, which means a total of 13/16 of the pizza had been consumed, and thus there was 3/16 of it left.

3. You will get twenty-one 1/4-kg servings of meat with 1/12 kg or 83 grams left over.
   5 1/3 ÷ (1/4) = (16/3) ÷ (1/4) = 16/3 × 4 = 64/3 = 21 1/3. So, you get 21 servings, and 1/3 of a serving.
   Since one serving is 1/4 kg, 1/3 of a serving is 1/12 kg.

4. Each piece is 1 4/9 ft long. 4 1/3 ÷ 3 = 13/3 × (1/3) = 13/9 = 1 4/9 ft.

5.

a. 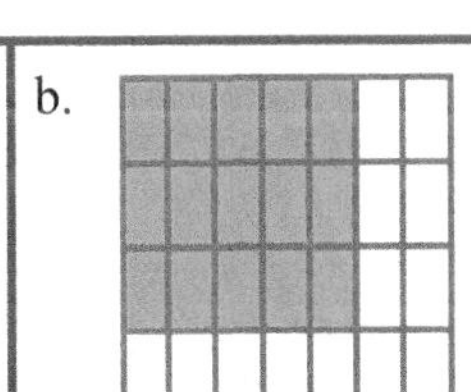

$\frac{2}{6} \times \frac{2}{3} = \frac{4}{18} = \frac{2}{9}$

b. $\frac{3}{4} \times \frac{5}{7} = \frac{15}{28}$

6.

a. $\frac{6}{7} \times \frac{5}{1} = \frac{30}{7} = 4\frac{2}{7}$	b. $\frac{12}{13} \times \frac{3}{7} = \frac{36}{91}$

7. The carpet costs $269.50.
   You can calculate the area of the room using either fractions or decimals. Using fractions, the area of the room is 11 ft × 8 3/4 ft = (11 ft × 8 ft) + (11 ft × 3/4 ft) = 88 sq. ft. + 33/4 sq. ft = 88 sq. ft + 8 1/4 sq. ft = 96 1/4 sq. ft.
   Using decimals, the area of the room is 11 ft × 8.75 ft = 96.25 sq. ft. To calculate the price, use decimal multiplication: 96.25 × $2.80 = $269.50

8. Answers will vary. Please check the students' answers. For example, there are 2 1/2 pizzas left, and three people share them evenly. How much pizza does each get? $2\frac{1}{2} \div 3 = \frac{5}{2} \times \frac{1}{3} = \frac{5}{6}$. Each person gets 5/6 of a pizza.

9. You can get six pieces with 1 1/2 ft left over.
   12 ft ÷ (1 3/4 ft) = 12 ÷ (7/4) = 12 × (4/7) = 48/7 = 6 6/7. This means you get 6 pieces and 6/7 of a piece. The six pieces are a total of 6 × (1 3/4 ft) = 6 18/4 ft = 10 1/2 ft, so since the string was 12 ft, you will have 1 1/2 ft of string left over.

10. Aiden now has $50.40.
    The ratio of 2:3 means Aiden got 3/5 of the reward. So, he got $120 ÷ 5 × 3 = $72. Then, since Aiden gave 3/10 of his money to his dad, he was left with 7/10 of it. $120 ÷ 10 × 7 = $50.40 (or you can also calculate it as $72 × 0.7).

Chapter 8 Test

Grading

My suggestion for grading is as follows. The total is 33 points. You can give partial points for partial solutions. Divide the student score by 33 to get a percent score. For example, if the student scores 22, divide 22 ÷ 33 with a calculator, getting 0.666666... The percent score is 67%.

Question #	Max. points	Student score
1	2 points	
2	4 points	
3	8 points	
4	4 points	

Question #	Max. points	Student score
5	9 points	
6	2 points	
7	4 points	
TOTAL	33 points	/ 33

1. $-5, -3, 0, 3$

2. a. $-7 + 2 = -5$

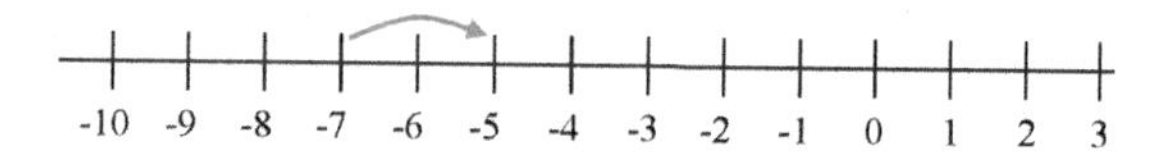

b. $-3 + 6 = 3$

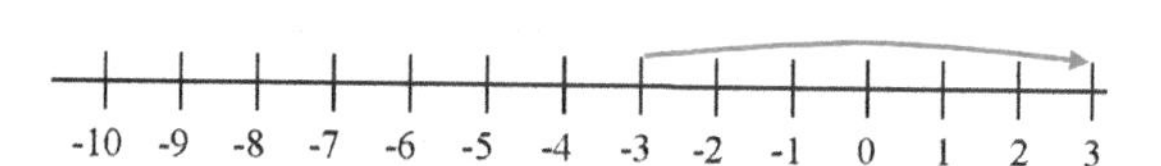

c. $-1 - 5 = -6$

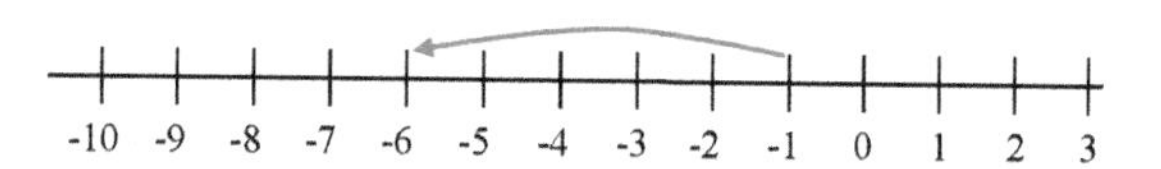

d. $2 - 7 = -5$

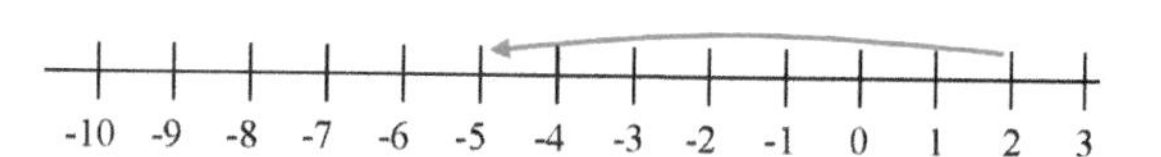

3.

a.	b.	c.	d.
$3 + (-7) = -4$	$(-1) + (-9) = -10$	$4 + (-5) = -1$	$8 - (-2) = 10$
$(-3) + 7 = 4$	$1 - 9 = -8$	$-4 - 5 = -9$	$-8 - (-2) = -6$

4. a. $|-9|$
   b. -43
   c. Henry's balance > −\$20. Students may also include that Henry's balance < 0, but it is not required. (The usual way to write these two statements is: −\$20 < Henry's balance < 0.)
   d. temperature < −10

5. a. Now her money situation is −\$11.

−\$3 − 8 = −\$11

b. Now the temperature is −3°C.

1°C − 4°C = −3°C

c. Now it is at the depth of −17 m.

−12 m + 5 m − 10 m = −17 m

6.

x	−7	−6	−5	−4	−3	−2	−1	0
y	−8	−7	−6	−5	−4	−3	−2	−1

x	1	2	3	4	5	6	7	8
y	0	1	2	3	4	5	6	7

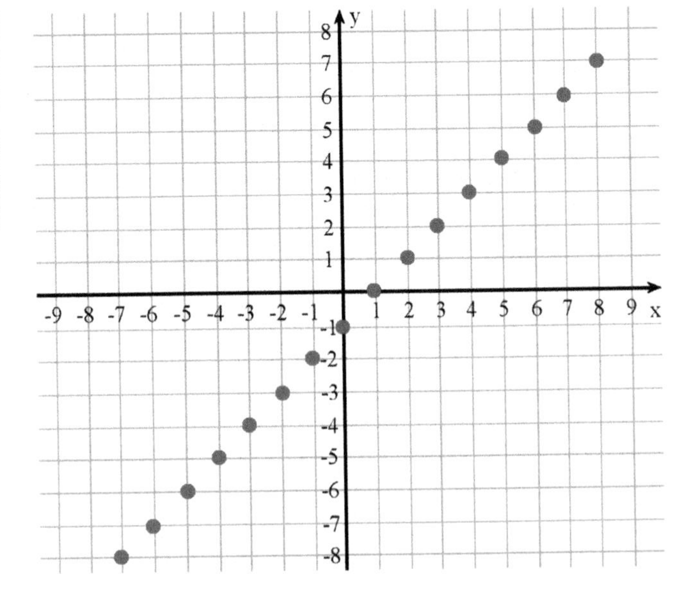

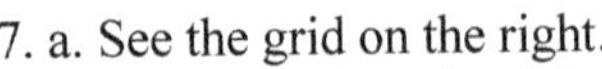

7. a. See the grid on the right.
   b. See the grid on the right.
   c. The new vertices are (−2, 6), (0, 2), and (4 , 4).

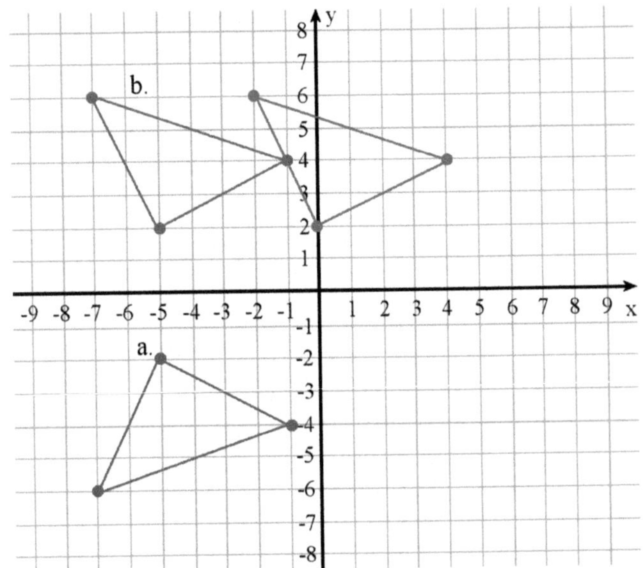

Chapter 9 Test

Grading

My suggestion for grading is as follows. The total is 26 points. You can give partial points for partial solutions. Divide the student's score by 26 to get a percent score. For example, if the student scores 21, divide 21 ÷ 26 with a calculator, getting 0.80769230... The percent score is 81%.

Question #	Max. points	Student score
1	4 points	
2	3 points	
3	4 points	
4	3 points	
5	2 points	

Question #	Max. points	Student score
6	2 points	
7	4 points	
8	2 points	
9	2 points	
TOTAL	26 points	/ 26

1. Check student's answers. The results may vary from those given below if the test was not printed at 100%.
   a. The area is 6.4 cm × 3.9 cm = 24.96 cm^2 ≈ 25 cm^2.
   b. 64 mm × 39 mm = 2,496 mm^2 ≈ 2,500 mm^2

2. It is 4 × 5 − 2 − 1 − 6 − 1.5 = 9.5 square units

3. a. trapezoid
   b. The area is 276 1/4 sq. in. The calculation for the area is (3 in × 13 in) ÷ 2 + (17 in × 13 in) + (5 1/2 in × 13 in) ÷ 2 = 19 1/2 sq. in. + 221 sq. in. + 35 3/4 sq. in. = 276 1/4 sq. in.

4. The front and back sides are 2 × 2 ft × 1.5 ft = 6 sq. ft. The two other sides are 2 × 1.5 ft × 1.5 ft = 4.5 sq. ft. The bottom is 2 ft × 1.5 ft = 3 sq. ft. The total area is 13.5 sq. ft.

5. The volume is 15/32 cubic inches. There are 30 little cubes, each having a volume of 1/64 cubic inches, so the total volume is 30/64 cubic inches = 15/32 cubic inches.

6. The volume is 19 1/2 cubic inches. Calculation: 6 1/2 in × 8 in × 3/8 in = 13/2 in × 8 in × 3/8 in = 52 × 3/8 in^3 = 13 × 3/2 in^3 = 19 1/2 in^3.

7. a. a square pyramid
   b. The surface area is 66 1/4 in^2.
   Area of the bottom: 5 in × 5 in = 25 in^2.
   One of the faces: 5 in × 4 1/8 in ÷ 2 = 5 in × 33/8 in ÷ 2 = 165/16 in^2 = 10 5/16 in^2.
   Total surface area: 25 in^2 + 4 × 10 5/16 in^2 = 25 in^2 + 41 1/4 in^2 = 66 1/4 in^2.

8. It is a triangular prism. Its net:

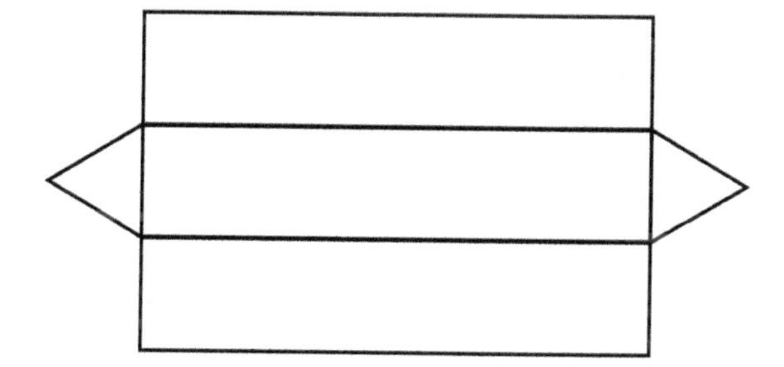

There are other possibilities, also.

9. The area is $4 \times 4 - 6 - 6 - 1/2 = 3\ 1/2$ square units

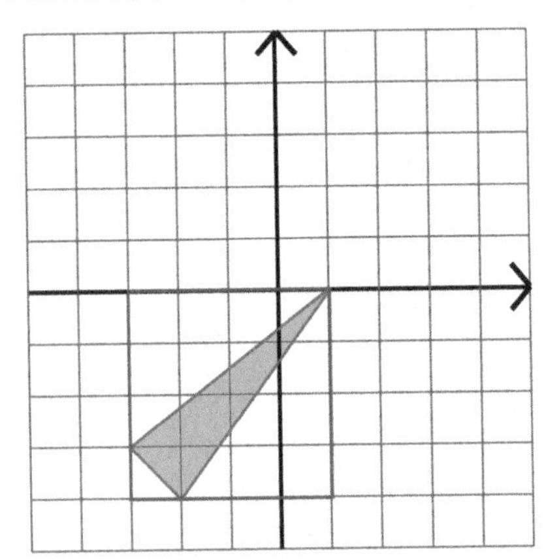

Chapter 10 Test

Grading

My suggestion for grading is as follows. The total is 25 points. You can give partial points for partial solutions. Multiply the student's score by 4 to get a percent score. For example, if the student scores 21, multiply 4 × 22 = 88. The percent score is 88%.

Question #	Max. points	Student score
1	8 points	
2a	2 points	
2b	2 points	
2c	2 points	
2d	1 point	

Question #	Max. points	Student score
3a	3 points	
3b	1 point	
3c	1 point	
4	5 points	
TOTAL	25 points	/ 25

1. a. mean 15.3 median 15 mode 15 range 9
   b. mean (none) median (none) mode horse, dog range (none)

2. a., b. and c.

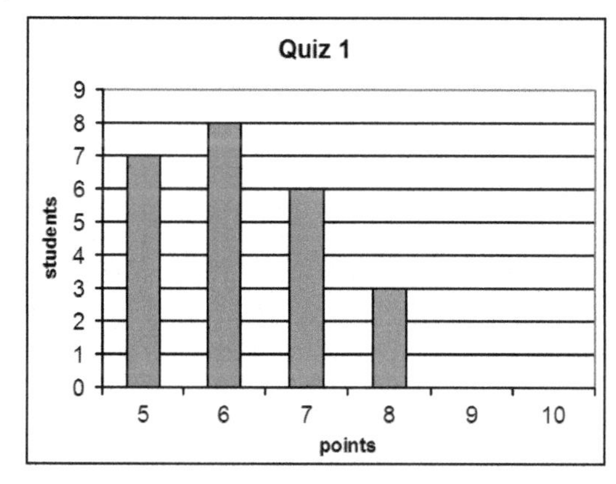

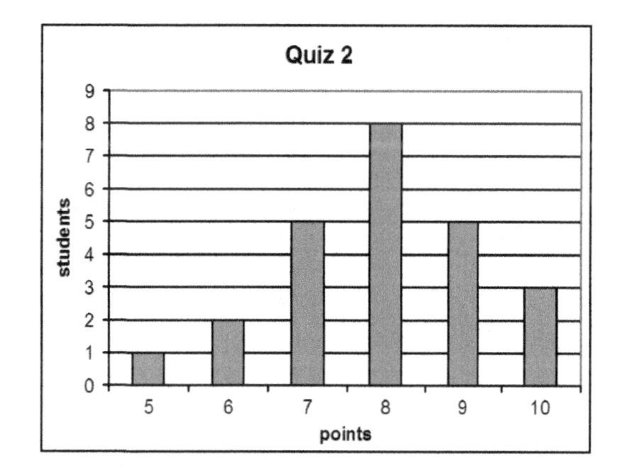

Quiz 1:

right-tailed or right-skewed

mean 6.21
median 6
mode 6

Quiz 2:

fairly bell-shaped

mean 7.96
median 8
mode 8

c. Any of the three measures of center can be used. All three are listed above.
d. Quiz 2

3. a.

Stem	Leaf
11	4 9
12	0 1 2 5 7 7 8
13	0 2

b. The median is 125.
c. The interquartile range is 128 − 120 = 8.

4.

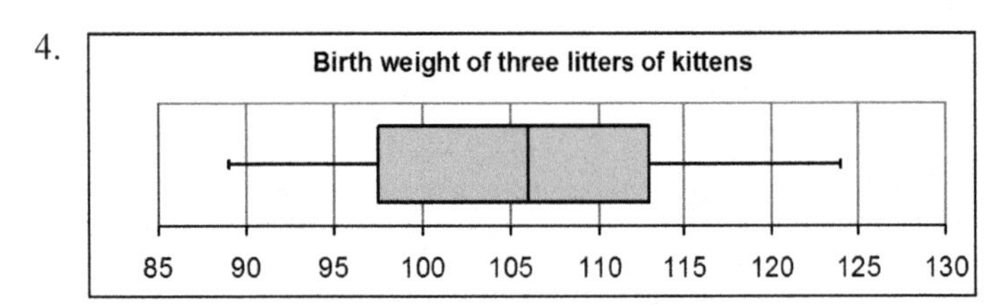

Five-number summary:
Minimum 89
1st quartile 97.5
Median 106
3rd quartile 113
Maximum 124

Math Mammoth End-of-the-Year Test, Grade 6, Answer Key

Instructions to the teacher:
In order to continue with the *Math Mammoth Grade 7 Complete Worktext*, I recommend that the student score a minimum of 80% on this test, and that the teacher or parent review with the student any content areas in which the student may be weak. Students scoring between 70% and 80% may also continue with grade 7, depending on the types of errors (careless errors or not remembering something, versus a lack of understanding). Use your judgment. My suggestion for points per item is as follows. The total is 194 points. A score of 155 points is 80%.

Question #	Max. points	Student score
Basic Operations		
1	2 points	
2	3 points	
3	2 points	
4	2 points	
	subtotal	/ 9
Expressions and Equations		
5	4 points	
6	2 points	
7	2 points	
8	1 point	
9	2 points	
10	2 points	
11	2 points	
12	2 points	
13	2 points	
14	2 points	
15	1 point	
16	2 points	
17	2 points	
18	2 points	
19	4 points	
	subtotal	/ 32
Decimals		
20	2 points	
21	2 points	
22	1 point	
23	2 points	
24	2 points	
25	1 point	
26	2 points	

Question #	Max. points	Student score
Decimals, cont.		
27	2 points	
28a	1 point	
28b	2 points	
29	3 points	
	subtotal	/ 20
Measuring Units		
30	3 points	
31	1 point	
32	2 points	
33	3 points	
34	6 points	
35	4 points	
	subtotal	/ 19
Ratio		
36	2 points	
37	2 points	
38	2 points	
39	2 points	
40	2 points	
41	2 points	
42	2 points	
	subtotal	/ 14
Percent		
43	3 points	
44	4 points	
45	2 points	
46	2 points	
47	2 points	
	subtotal	/13

Question #	Max. points	Student score
Prime Factorization, GCF, and LCM		
48	3 points	
49	2 points	
50	2 points	
51	2 points	
52	2 points	
	subtotal	/11
Fractions		
53	3 points	
54	2 points	
55	2 points	
56	2 points	
57	3 points	
58	3 points	
	subtotal	/15
Integers		
59	2 points	
60	2 points	
61	2 points	
62	4 points	
63	5 points	
64	6 points	
65	4 points	
	subtotal	/25

Question #	Max. points	Student score
Geometry		
66	1 point	
67	1 point	
68	3 points	
69	4 points	
70	2 points	
71a	1 point	
71b	3 points	
72	4 points	
73a	2 points	
73b	2 points	
	subtotal	/23
Statistics		
74a	2 points	
74b	1 point	
74c	2 points	
75a	1 point	
75b	1 point	
76a	2 points	
76b	1 point	
76c	1 point	
76d	2 points	
	subtotal	/13
	TOTAL	/194

The Basic Operations

1. a. 2,000 ÷ 38 = 52 R4. There will be 52 bags of cinnamon.

2. a. $2^5 = 32$ b. $5^3 = 125$ c. $10^7 = 10{,}000{,}000$

3. a. 70,200,009
   b. 304,500,100

4. a. 6,300,000
   b. 6,609,900

Expressions and Equations

5. a. $s - 2$ b. $(7 + x)^2$ c. $5(y - 2)$ d. $\dfrac{4}{x^2}$

6. a. $40 - 16 = 24$

   b. $\dfrac{65}{5} = 13 \cdot 3 = 39$

7. a. $\$50 - 2m$ or $\$50 - m \cdot 2$
   b. s^2

8. $z + z + 8 + x + x + x = 2z + 3x + 8$ or $3x + 2z + 8$ or $2z + 8 + 3x$

9. $6(s + 6)$ or $(s + 6 + s + 6 + s + 6 + s + 6 + s + 6 + s + 6$. It simplifies to $6s + 36$.

10. $6b \cdot 3b = 18b^2$

11. a. $3x$ b. $14w^3$

12. a. $7(x + 5) = 7x + 35$
    b. $2(6p + 5) = 12p + 10$

13. a. $\underline{2}(6x + 5) = 12x + 10$
    b. $5(2h + \underline{6}) = 10h + 30$

14.

a.	b.
$\dfrac{x}{31} = 6$	$a - 8.1 = 2.8$
$x = 6 \cdot 31$	$a = 2.8 + 8.1$
$x = 186$	$a = 10.9$

15. $y = 2$

16. $0.25 \cdot x = 16.75$ OR $25x = 1675$. The solution is $x = 67$ quarters.

17. a. $p \le 5$. The variable students use for "pieces of bread" may vary.
    b. $a \ge 21$. The variable students use for "age" may vary.

18.

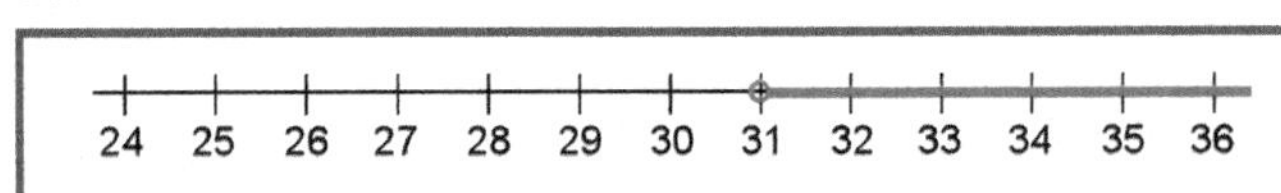

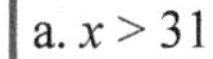
a. $x > 31$

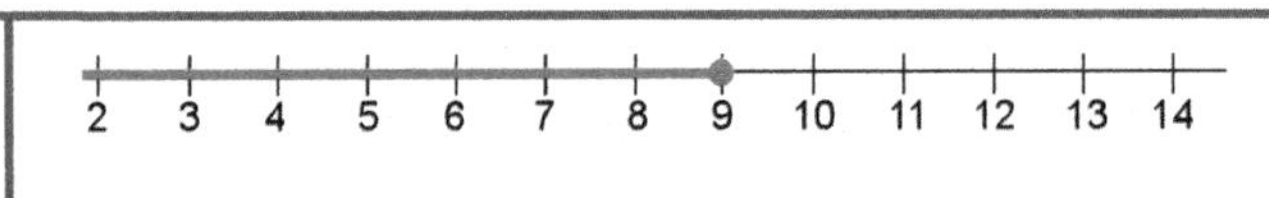

b. $x \le 9$

19. a.

t (hours)	0	1	2	3	4	5	6
d (km)	0	80	160	240	320	400	480

b. See the grid on the right.
c. $d = 80t$
d. t is the independent variable

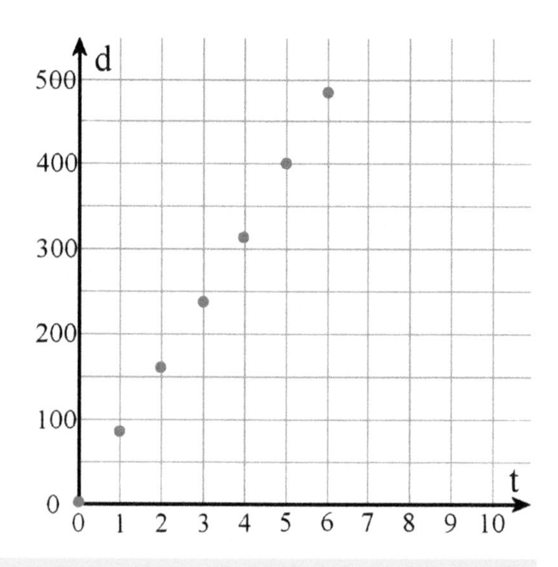

Decimals

20. a. 0.000013 b. 2.0928

21. a. $\frac{78}{100,000}$ b. $2\ \frac{302}{1,000,000}$

22. 0.0702

23. a. 8
b. 0.00048

24. a. Estimate: $7 \times 0.006 = 0.042$
b. Exact: $7.1 \times 0.0058 = 0.04118$

25. $1.5 + 0.0022 = 1.5022$

26. a. 90,500
b. 0.0024

27. a. $175 \div 0.3 = 583.333$

b. $\frac{2}{9} = 0.222$

28. a. Estimate: $13 \div 4 \times 3 = (3\ 1/4) \times 3 = \$9.75$
b. Exact: $9.69

29. $(3 \times \$3.85 + \$4.56) \div 2 = \$8.06$

Measuring Units

30. a. 178 fl. oz. = 5.56 qt b. 0.412 mi. = 2,175.36 ft c. 1.267 lb = 20.27 oz

31. 0.947 mile

32. You can get 10 six ounce serving and have 4 ounces left over.

33. It is about $6.65 per pound.
To calculate the price per pound, simply divide the cost by the weight in pounds. A pack of 36 candy bars weighs 36×1.55 oz = 55.8 oz = 3.4875 lb. Now simply divide the cost of those candy bars by their weight in pounds to get the price per pound: $23.20 ÷ 3.4875 lb = $6.652329749103943 / lb.

34. a. 39 dl = 3.9 L

			3	9		
kl	hl	dal	l	dl	cl	ml

b. 15,400 mm = 15.4 m

		1	5	4	0	0
km	hm	dam	m	dm	cm	mm

c. 7.5 hm = 75,000 cm

	7	5	0	0	0	
km	hm	dam	m	dm	cm	mm

d. 597 hl = 59,700 L

5	9	7	0	0			
	kl	hl	dal	l	dl	cl	ml

e. 7.5 hg = 0.75 kg

0	7	5				
kg	hg	dag	g	dg	cg	mg

f. 32 g = 3,200 cg

		3	2	0	0	
kg	hg	dag	g	dg	cg	mg

35. a. Twenty-four bricks will cover the span of the wall. 5150 mm ÷ 215 mm = 23.953488.
b. Twenty-three bricks will cover the span of the wall. 5150 mm ÷ 225 mm = 22.88

Ratio

36. a.

b. 10:15 = 2:3

37. a. 3,000 g:800 g = 15:4
b. 240 cm:100 cm = 12:5

38. a. $7:2 kg
b. 1 teacher per 18 students

39. a. $4 per t-shirt.
b. 90 miles in an hour

40. a. You could mow 20 lawns in 35 hours.
b. The unit rate is 105 minutes per lawn (or 1 h 45 min per lawn).

Lawns	4	8	12	16	20
Hours	7	14	21	28	35

41. Mick got $102.84. $180 ÷ 7 × 4 = $102.84.

42. a. 11.394 km b. 4.23 qt

Percent

43.

a. 35% = $\frac{35}{100}$ = 0.35	b. 9% = $\frac{9}{100}$ = 0.09	c. 105% = $1\frac{5}{100}$ = 1.05

44.

	510
1% of the number	5.1
5% of the number	25.5
10% of the number	51
30% of the number	153

45. The discounted price is $39. You can multiply 0.6 × $65 = $39, or you can find out 10% of the price, which is $6.50, multiply that by 4 to get the discount ($26), and subtract the discounted amount.

46. The store had 450 notebooks at first. Since 90 is 1/5 of the notebooks, the total is 90 ×5 = 450.

47. She has read 85% of the books she borrowed from the library. 17/20 = 85/100 = 85%.

Prime Factorization, GCF, and LCM

48. a. 3 × 3 × 5 b. 2 × 3 × 13 c. 97 is a prime number

49. a. 8 b. 18

50. a. 2 b. 15

51. Any three of the following numbers will work: 112, 140, 168, 196

52.

a. GCF of 18 and 21 is 3. 18 + 21 = 3· 6 + 3 ·7 = 3(6 + 7)
b. GCF of 56 and 35 is 7. 56 + 35 = 7(8 + 5)

Fractions

53. a. 4 b. 2 1/12 c. 5 3/5

54. $3\frac{2}{3} \div \frac{3}{5} = 6\frac{1}{9}$

55. Answers will vary. Please check the student's work.
Example: There was 1 3/4 pizza left over and three people shared it equally. Each person got 7/12 of a pizza.

56. There are ten servings. (7 1/2) ÷ (3/4) = (15/2) ÷ (3/4) = (15/2) × (4/3) = 60/6 = 10.

57. 63 8/9 square feet.
The area of the room is (12 1/2) × (15 1/3) = (25/2) × (46/3) = 25 × 23/3 = 575/3 = 191 2/3 square feet.
One-third of that is (191 2/3) × (1/3) = 574/9 = 63 8/9.
Or, you can first divide one of the dimensions by three, and then multiply to find the area.

58. 4 13/20 inches and 3 1/10 inches or 4.65 inches and 3.1 inches.

The ratio of 3:2 means the two sides are as if three "parts" and two "parts", and the total perimeter is 10 of those parts. Therefore, one part is 15 1/2 in. ÷ 10 = 15.5 in. ÷ 10 = 1.55 inches. The one side is three times that, and the other is two times that. So, the sides are 4.65 in. and 3.1 in. If you use fractions, you get (15 1/2 in.) ÷ 10 = (31/2 in.) ÷ 10 = 31/20 in., and the two sides are then 3 × 31/20 in. = 93/20 in. = 4 13/20 in. and 2 × 31/20 in. = 62/20 in. = 3 1/10 in.

Integers

59. a. > b. >

60. a. −7°C > −12°C.
    b. \$5 > −\$5.

61. a. The difference is 23 degrees.
    b. The difference is 12 degrees.

62. a. −7 b. $|-6| = 6$ c. $|5| = 5$ d. $|-6| = 6$

63. a.- c See the grid on the right.
    d. $6 \times 10 \div 2 = 30$
    The area of the resulting triangle is 30 square units.

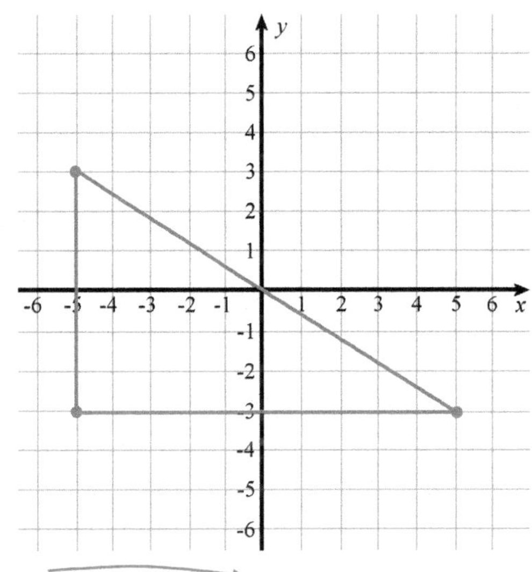

64. a. $-2 + 5 = 3$

b. $-2 - 4 = -6$

c. $-1 - 5 = -6$

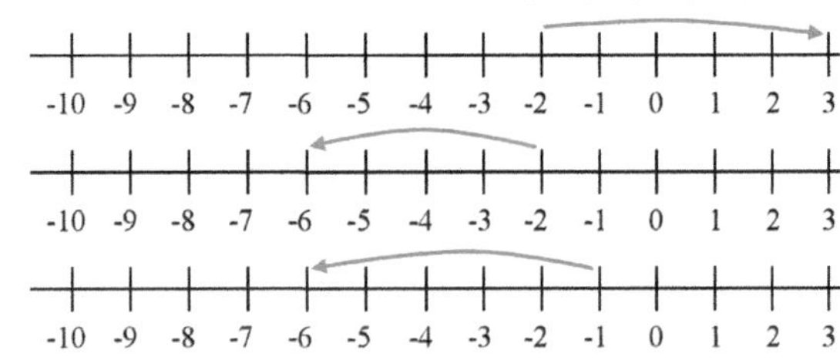

65. a. That would make his money situation to be −\$4.

\$10 −\$14 = −\$4
OR
\$10 + (−\$14) = −\$4

b. Now he is at the depth of −3 m.

−2 m −1 m = −3 m
OR
−2 m + (−1 m) = −3 m

Geometry

66. 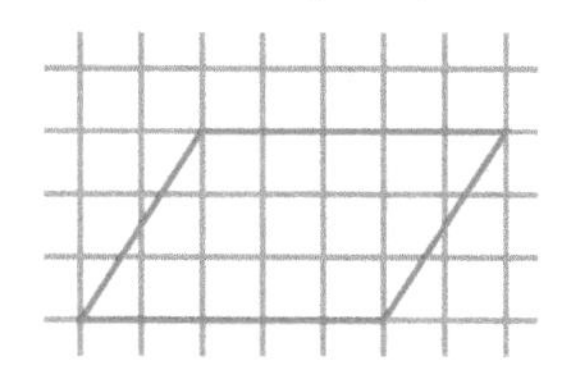 The area is $4 \times 3 \div 2 = 6$ square units.

67. Answers may vary. The base and altitude of the parallelogram could be for example 5 and 3, or 3 and 5, or 6 and 2 1/2.

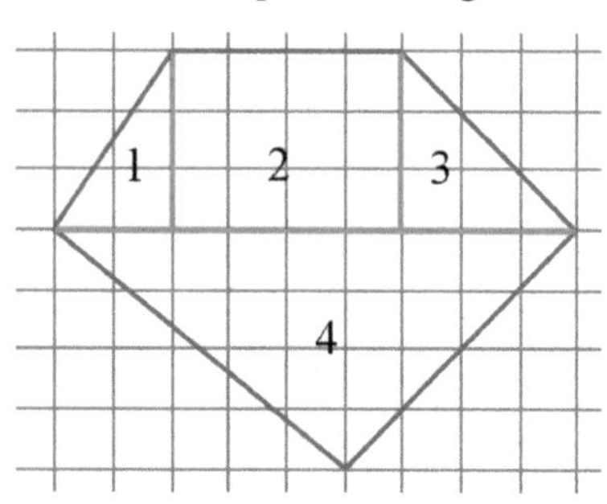

68. Divide the shape into triangles and rectangles, for example like this:

The areas of the parts are:

triangle 1: 3 square units
rectangle 2: 12 square units
triangle 3: 4.5 square units
triangle 4: 18 square units

The overall shape (pentagon): 37.5 square units

69. It is a trapezoid. To calculate its area, divide it into triangles and rectangle(s).

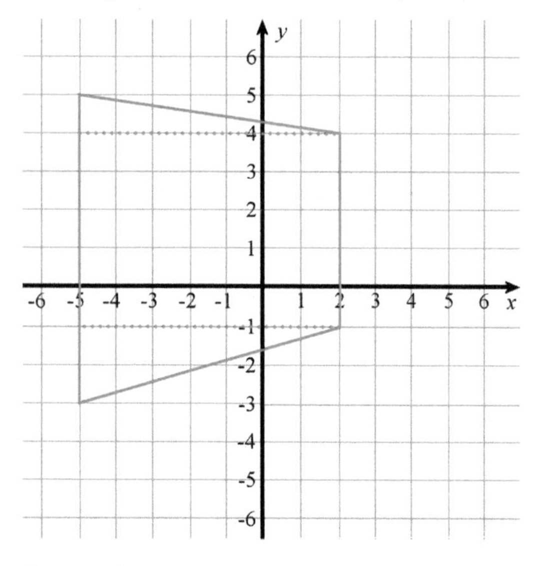

The area is: 3.5 + 35 + 7 = 45.5 square units

70. It is a triangular prism. Some possible nets are shown below:

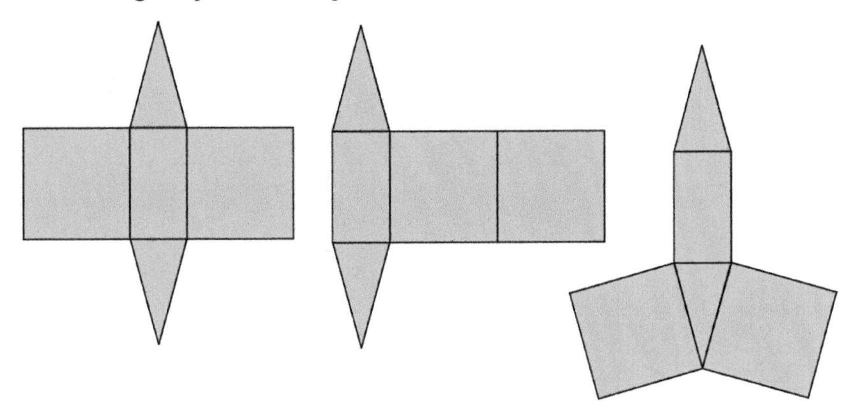

71. a. It is a rectangular pyramid.

b. The rectangle has the area of 300 cm². The top and bottom triangles: 2 × 20 cm × 11.2 cm ÷ 2 = 224 cm². The left and right triangles: 2 × 15 cm × 13 cm ÷ 2 = 195 cm². The total surface area is 719 cm².

72. The volume of each little cube is (1/2 cm) × (1/2 cm) × (1/2 cm) = 1/8 cm³.
a. 18 × (1/8) cm³ = 18/8 cm³ = 9/4 cm³ = 2 1/4 cm³.
b. 36 × (1/8) cm³ = 36/8 cm³ = 9/2 cm³ = 4 1/2 cm³.

73. a. 1 3/4 in × 8 1/2 in × 6 in = (7/4) in × (17/2) in × 6 in = (119/4) × 6 in³ = (29 3/4) × 3 in³ = 87 9/4 in³ = 89 1/4 in³. This calculation can also be done (probably quicker) by using decimals: 1.75 in × 8.5 in × 6 in = 89.25 in³.

b. Imagine you place the boxes in rows, standing up, so that the height is 6 inches. Then we can stack two rows on top of each other, since the height of the box is 1 ft or 12 inches. The width of each box is 1 3/4 in., and 6 boxes fit in the space of 1 ft., because 6 × (1 3/4 in.) = 6 18/4 in. = 10 1/2 in. Since the last dimension is over 8 inches, we cannot fit but one row. So, we can fit two rows of 6 boxes, stacked on top of each other, or a total of 12 boxes.

Statistics

74. a. See the plot on the right.
b. The median is 68.5 years.
c. The first quartile is 63, and the third quartile is 75.5. The interquartile range is thus 12.5 years.

Stem	Leaf
5	5 9
6	1 2 4 5 5 8 9
7	0 2 4 7
8	3 9
9	4

75. a. It is right-tailed or right-skewed. You can also describe it as asymmetrical.
b. Median. Mean is definitely not the best, because the distribution is so skewed. Without seeing the data itself, we cannot know if mode would work or not - it may not even exist, since typically for histograms, the data is very varied numerically and has to first be grouped.

76. a.

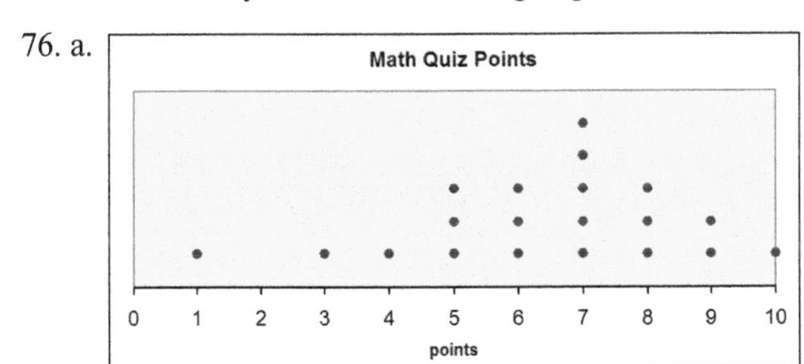

b. It is fairly bell-shaped but is somewhat left-tailed or left-skewed. You can also say it is asymmetrical.
c. The data is spread out a lot.
d. Any of the three measures of center works. Mean: 6.4. Median: 7. Mode: 7.

Cumulative Reviews Answer Key

Cumulative Reviews Answer Key, Grade 6

Cumulative Review: Chapters 1 - 2

1\. a.

Distance	4 km	12 km	20 km	24 km	132 km	216 km
Tim	10 min	30 min	50 min	1 hour	5 1/2 hours	9 hours

b. To travel 360 kilometers will take 360 km ÷ 24 km/hr = 15 hours.

2\. a. Answers will vary. Check the student's answers. The simplest estimate is to round 1,091 down to 1,000:
234 · 1,000 = 234,000. (You might compensate for rounding one number down by rounding the other one up:
240 · 1,000 = 240,000.)
b. 234 · 1,091 = 255,294
c. Answers will vary. Check the student's answers. The simplest estimate is to round 1.091 down to 1: 2.34 · 1 = 2.34.
(You might compensate for rounding one number down by rounding the other one up: 2.4 · 1 = 2.4.)
d. Adjusting the decimal points in our estimate from part (a), we can see that 2.34 · 1 is 2.34, so we also know that
2.34 · 1.091 is about that same amount, so, based on part (b), the exact answer must be 2.55294.

Enrichment (optional): Exercise #2 is a fascinating problem because it lends itself so well to what is called an "iterative" (*i.e.*, "in steps") method. Although the simplest estimate simply rounds 1,091 down to 1,000, that estimate will be too small because of rounding down. We could get a little closer by recognizing that 91 is almost 100. (We are rounding 91 up, so this estimate will be too big.) Since 234 · 100 = 23,400, a closer estimate would be 234 · 1,100 = 234 · 1,000 + 234 · 100 = 234,000 + 23,400 = 257,400. The exact answer is somewhere between 234,000 and 257,400, but a lot (91/100) closer to 257,400. We can even keep going, subtracting 234 · 10 = 2,340 to get 234 · 1,090 = 257,400 − 2,340 ≈ 255,000, which gets us accuracy to the nearest thousand. In other words, 234 · 1,091 = 234 · 1,000 + 234 · 100 − 234 · 10 + 234 · 1, so we can transform the multiplication problem into a simpler addition/subtraction problem and estimate the answer to whatever place value we want.

3\. a. $56 - y = 17$ (*Minuend − subtrahend = difference.*) Solution: $y = 56 - 17 = 39$.
b. $x \div 15 = 60$ (*Dividend ÷ divisor = quotient.*) Solution: $x = 60 \cdot 15 = 900$.

4\. a. 600 ÷ 6 + 36 ÷ 6 = 100 + 6 = 106
b. 800 ÷ 4 + 24 ÷ 4 = 200 + 6 = 206
c. 5,600 ÷ 7 + 7 ÷ 7 = 800 + 1 = 801
d. 1,200 ÷ 12 + 24 ÷ 12 = 100 + 2 = 102

5\. a. 100 − (100 ÷ 4) · 2 = 100 − (25 · 2) = 100 − 50 = 50
b. $3^3 \div (4 + 5) = 27 \div 9 = 3$
c. $(2 + 6)^2 - (25 - 5) = 8^2 - 20 = 64 - 20 = 44$
d. (144 + 9)/(5 · 3) = 153/15 = 10 1/5

6\. a. 3 · 5 − 12 = 15 − 12 = 3
b. 24/3 + 4 = 8 + 4 = 12

7\. The perimeter is the sum of the lengths of the sides, or $l + w + l + w$, so the correct answer is (b) $2l + 2w$.
(Answer (a) gives the rectangle's area. Answer (c) gives the aspect ratio. Answers (d) and (e) give partial perimeters.)

8\. a. $A = (11\text{ cm})^2 = 121\text{ cm}^2$
b. $V = (4\text{ ft})^3 = 64\text{ ft}^3$

9\. To find the length of one side of the square divide 64 cm by 4. Then the area would be $A = (16\text{ cm})^2 = 256\text{ cm}^2$.

10\. $48.60/3 · 2 = $16.20 · 2 = $32.40. The total cost is $48.60 + $32.40 = $81. Her change is $19.00.

11\. a. 15,711
b. 0.533
c. 0.043

Cumulative Review: Chapters 1 - 3

1. One parent paid \$103.74 and the other three parents paid \$34.58 each.

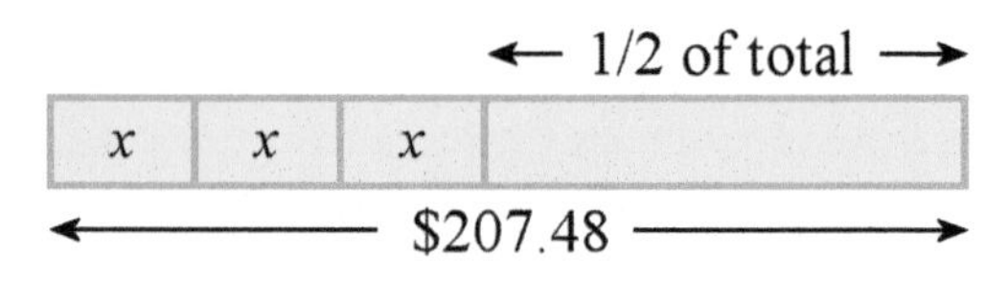

2. a. 329,000,300
   b. 1,050,003

3. a. 5,700,000
   b. 219,997,000
   c. 83,000,000
   d. 3,999,990,000

4.

Variable	Expression $\frac{x^2}{3}$	Value
$x = 1$	$\frac{1^2}{3}$	$\frac{1}{3}$
$x = 2$	$\frac{2^2}{3} = \frac{4}{3}$	$1\frac{1}{3}$

Variable	Expression $\frac{x^2}{3}$	Value
$x = 3$	$\frac{3^2}{3}$	3
$x = 5$	$\frac{5^2}{3} = \frac{25}{3}$	$8\frac{1}{3}$

5. a. $(12 + 56) \div 4$ or $\frac{12 + 56}{4} = 68 \div 4 = 17$.

   b. $8 \div 4^3$ or $\frac{8}{4^3} = \frac{1}{8}$.

6. a. $8c^4$
   b. $5c + 8$
   c. $t + 3$
   d. $13x^2 + 13$

7. a. Her friend got $(1/3)m$ or $m/3$.
   b. Fanny is $s - 6$.
   c. Sadie will be $s + 5$.
   d. Fanny will be $s - 1$.

8.

a. $7x + 2x = 54$ $9x = 54$ $x = 54 \div 9$ $x = 6$	b. $8r - 3r = 40$ $5r = 40$ $r = 40 \div 5$ $r = 8$	c. $t \div 50 = 5 + 6$ $t \div 50 = 11$ $t = 50 \cdot 11$ $t = 550$
d. $w - 88 = 20 \cdot 60$ $w - 88 = 1{,}200$ $w = 1{,}200 + 88$ $w = 1{,}288$	e. $2x - 6 = 16$ $2x = 16 + 6$ $x = 22 \div 2$ $x = 11$	f. $8x + 17 = 81$ $8x = 81 - 17$ $x = 64 \div 8$ $x = 8$

9.

a. $16y + 12 = 4(4y + 3)$	b. $9x + 9 = 9(x + 1)$
c. $54c + 24 = 6(9c + 4)$	d. $15a + 45 = 15(a + 3)$

10. a. $x = 7 - 4.5039 = 2.4961$
    b. $x = 0.938208 - 0.047 = 0.891208$
    c. $x = 6.0184 \div 2 = 3.0092$

Cumulative Review: Chapters 1 - 4

1. a. Equation: $x \div 11 = 12$ or $\frac{x}{11} = 12$. Solution: $x = 11 \cdot 12 = 132$.

   b. Equation: $3 + 5 + x = 105$. Solution: $x = 105 - 5 - 3 = 97$.

2.

a.	b.	c.
$x \div 6 = 40 + 50$	$1{,}000 - x = 40 \cdot 6$	$8x + 2x = 15 \cdot 6$
$x \div 6 = 90$	$1{,}000 - x = 240$	$10x = 90$
$x = 90 \cdot 6$	$1{,}000 = 240 + x$	$x = 90 \div 10$
$x = 540$	$1000 - 240 = x$	$x = 9$
	$760 = x$	

3. a. The average price is the sum of the values divided by the number of items:
   ($3.89 + $3.99 + $4.45 + $3.79 + $4.10 + $4.19 + $4.02) / 7 = $28.43 / 7 = $4.06
   b. She saves $445 − $379 = $66.

4.

a. $10 \cdot 0.009 = 0.09$ $0.5 \cdot 0.6 = 0.3$	b. $40 \cdot 0.08 = 3.2$ $1{,}000 \cdot 1.2 = 1200$	c. $0.1 \cdot 0.2 \cdot 0.3 = 0.006$ $0.11 \cdot 0.02 = 0.0022$
d. $10 \div 0.2 = 50$ $0.6 \div 0.2 = 3$	e. $0.075 \div 0.025 = 3$ $0.3 \div 0.02 = 15$	f. $2.36 \div 2 = 1.18$ $0.0045 \div 5 = 0.0009$

5.

a. 6 kg = <u>6,000</u> g 5 dl = <u>0.5</u> L 5 mm = <u>0.005</u> m	b. 7 dam = <u>70</u> m 5 hl = <u>500</u> L 30 cg = <u>0.3</u> g	c. 7 kl = <u>7,000</u> L 50 mg = <u>0.05</u> g 8 cm = <u>0.08</u> m

6. a. Tim's weight compared to the grasshopper's is 45,000 g : 3 g, so <u>Tim weighs 15,000 times more</u>.
   b. You could easily carry the weight of a thousand grasshoppers because it would be only 1000 · 3 g = 3 kg (about 6 ½ lb).

7. According to the graphic, 5/6 of Elaine's after-tax salary (in brown) was $1,000, so each sixth (brown part) was $1,000 ÷ 5 = $200, and her total after-tax salary was 6 · $200 = $1,200. But her after-tax salary was 4/5 of her total salary (in blue), so each fifth of her total salary was $1,200 ÷ 4 = $300. So, her total salary was 5 · $300 = <u>$1,500</u>.

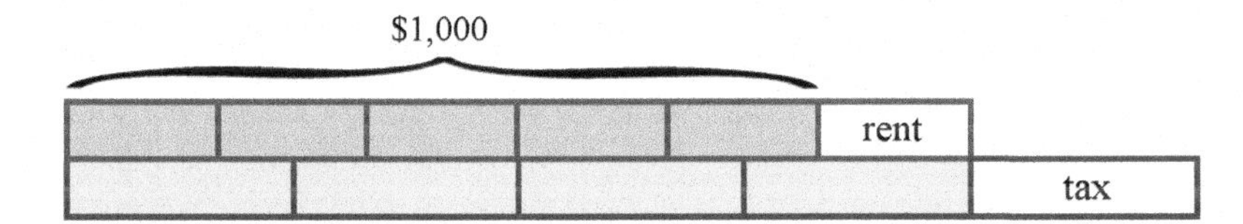

8. a. $45.7 \div 0.02 = 2{,}285$
   b. $928 \div 0.003 \approx 309{,}333.33$
   c. $\frac{5}{8} = 0.625$

Cumulative Review: Chapters 1 - 5

1. a. 0.00392 b. 5.0015
   c. 0.000023 d. 12.012

2. a. 16/1,000,000
   b. 29381/10,000
   c. 39,402/100,000

3. a. $1 - 0.05 = 0.95$
   b. $0.1 - 0.05 = 0.05$
   c. $1.1 - 0.05 = 1.05$

4.

	2.97167	0.046394	2.33999	1.199593
the nearest tenth	3.0	0.0	2.3	1.2
the nearest thousandth	2.972	0.046	2.340	1.200

5. Final answers do not vary, but the ways to get there can vary.

a.	b.	c.
$\frac{5.6}{0.4} = \frac{56}{4} = 14$	$\frac{4}{0.02} = \frac{40}{0.2} = \frac{400}{2} = 200$	$\frac{0.9}{0.003} = \frac{9}{0.03} = \frac{90}{0.3} = \frac{900}{3} = 300$

6. $\frac{320 \text{ people}}{1{,}200 \text{ people}} = \frac{4}{15} = \frac{40 \text{ people}}{150 \text{ people}}$

   a. The ratio of people who like mashed potatoes best to the total number of people interviewed is <u>4:15</u>.
   b. Of a group of 150 people, we would expect <u>about 40</u> to prefer mashed potatoes.

7. a. $\frac{14 \text{ km}}{20 \text{ min}} = \frac{3.5 \text{ km}}{5 \text{ min}} = \frac{31.5 \text{ km}}{45 \text{ min}}$ b. $\frac{\$33.60}{8 \text{ bottles}} = \frac{\$4.20}{1 \text{ bottle}} = \frac{\$42}{10 \text{ bottles}}$

8. $\frac{2 \text{ kg}}{120 \text{ m}^2} = \frac{1 \text{ kg}}{60 \text{ m}^2} = \frac{5 \text{ kg}}{300 \text{ m}^2}$ The lawn has an area of $15 \text{ m} \cdot 20 \text{ m} = 300 \text{ m}^2$, so you would need <u>5 kg</u> of fertilizer.

9.

x	2	3	4	5	6	7
y	0	2	4	6	8	10

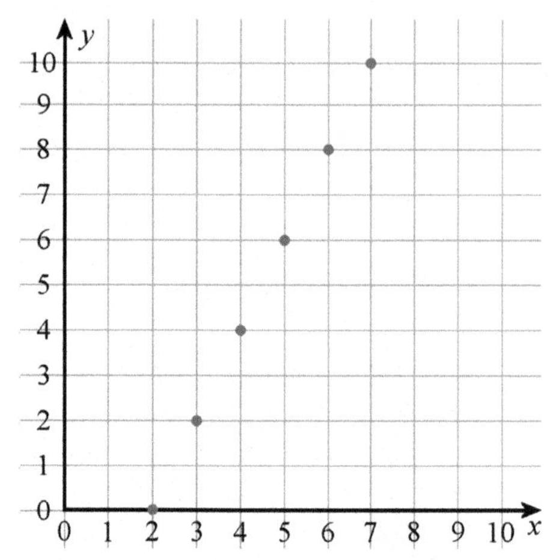

10. If two-thirds of a stick is 50 cm long, then each third is 25 cm long, and the whole stick is three-thirds long, or $3 \cdot 25$ cm = <u>75 cm</u>.

11. Two gallons is 256 ounces. Then, 256 oz ÷ 6 oz = 42 servings with 4 oz left over.

12. a. Hannah 151 cm; Erica 136 cm
    b. Hannah 175 cm or 1,750 mm; Erica 160 cm or 1,600 mm.
    c. Hannah 165 cm or 1.65 m; Erica 150 cm or 1.5 m.

13. Every afternoon Erica bicycles 5 miles (<u>8.0 km</u>) to the horse ranch. Erica takes care of a horse that is 15 *hands*, or 60 inches (<u>1.5 m</u>), tall. She likes to go riding on a trail that is 4 mi 500 ft (<u>6.6 km</u>) long.

(5 mi · 1.6093 km/mi = 8.0465 km)
(60 in · 2.54 cm/in = 152.4 cm)
(21,620 ft · 0.3048 m/ft ≈ 6,590 m)

Cumulative Review: Chapters 1 - 6

1. a. 80%   b. 85%   c. 45%

2. a. 0.8   b. 0.85   c. 0.45

3. a. 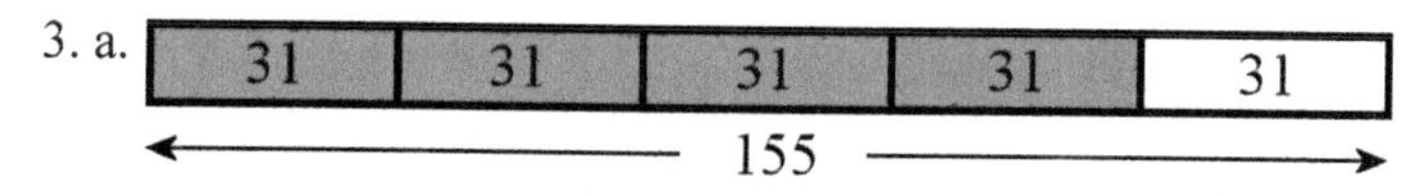

   b. 4/5 were basic calculators
   c. 80% were basic calculators
   d. There were 31 scientific calculators.

4.

Distance	2 km	4 km	5 km	16 km	20 km	24 km	40 km	60 km	70 km
Time	6 min	12 min	15 min	48 min	1 hour	1 h 12 min	2 hours	3 hours	3 1/2 hours

5. $53.75
   The price of the other flash drive is $25 + $2.50 + $1.25 = $28.75. Buying both, the cost is $25 + $28.75 = $53.75.

6.

a. $32t + 8 = 8(4t + 1)$	b. $8 + 12x = 4(2 + 3x)$
c. $15y + 45 = 3(5y + 15)$	d. $35 + 42w = 7(5 + 6w)$

7. Her score is 35/40 = 7/8 = 87.5% .

8. He has 9 cars left. One-fifth of his cars is 18 cars, so that is how many Jack kept. Then he gave half of those to his brother.

9.

a. $10{,}000 \times 0.092 = 920$	b. $1{,}000 \times 0.0004 = 0.4$
c. $456.29 \div 1{,}000 = 0.45629$	d. $63 \div 10^5 = 0.00063$

10. a. 15.34 m   b. 0.334 L   c. 900 g

11. a. You need to first choose a variable for the unknown. Let b be the cost of the camera bag.
    The equation is $b + \$85 = \$162$ or $\$162 - \$85 = b$. The camera bag costs $77.
    b. Let p be the price of one towel. $8t = \$52$. One towel cost $6.50.

12. Three shirts cost $32.90. $14.10 ÷ 3 × 7 = $32.90

13.

a. 79 oz = 4 lb 15 oz b. 4 ft 11 in = 59 in	c. 7.82 qt = 1.96 gal d. 0.265 mi = 466.4 yd	e. 2.54 lb = 40.64 oz f. 6.8 ft = 6 ft 9.6 in

Cumulative Review: Chapters 1 - 7

1. a. 8   b. 18

2. a. 1   b. 12

3.

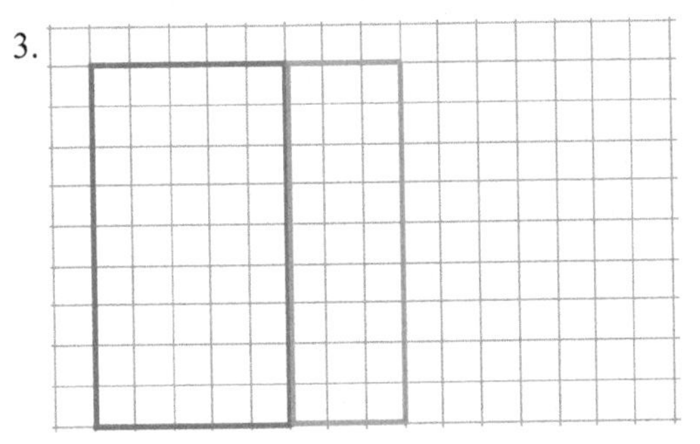

4. Each person gets 7.11 ounces of ice cream. Two quarts is 64 ounces, so each person gets $2 \times 32 \div 9 = 7.11$ oz.

5. a. 11^2 gives us the area of a square with a side length of 11 units.
   b. 3×5^2 gives us the area of 3 squares with a side length of 5 units.
   c. 4×0.4^3 gives us the volume of 4 cubes with an edge length of 0.4 units.

6. a. 26,020,000
   b. 1,000,200,807

7. a. $3 \times 3 \times 11$
   b. $2 \times 2 \times 2 \times 2 \times 7$
   c. $2 \times 2 \times 2 \times 5 \times 5$

8. The total capacity is 49.5 liters.
   The smaller container holds $8.5 \times 0.60 = 5.1$ liters. The total capacity is then $3 \times 8.5 + 4 \times 5.1 = 45.9$ liters.

9. Samantha earned \$25 more than George. Samantha earned $\$100 \div 8 \times 5 = \$62.50$ and George earned \$37.50. Or, you can think this way: Samantha earned 5 parts and George 3 parts; therefore the difference in their earnings is 2 parts. Each part is $\$100 \div 8 = \$12.50$, so two parts is \$25.

10. a. $\frac{6}{7s}$ b. $11 - 2x$
    c. $(x + 2)^2$ d. $(5m)^3$
    e. $\frac{2t^2}{s - 1}$ f. $18 - y$

11. $2x + 194 = 388$
    $2x = 388 - 194$
    $x = 97$

x	x	194

← 388 →

12.

a. $4 \times 0.7 = 2.8$ b. $50 \times 0.003 = 0.15$	c. $3 \times 1.06 = 3.18$ d. $100 \times 0.009 = 0.9$	e. $10^5 \times 0.08 = 8{,}000$ f. $40 \times 0.004 = 0.16$

Cumulative Review: Chapters 1 - 8

1.

a. $\frac{23}{24}$	b. $\frac{11}{35}$	c. $\frac{1}{79}$	d. $\frac{1}{100}$	e. $\frac{1000}{3}$

2.

a. $\frac{6}{7} \div \frac{1}{7} = 6$	b. $\frac{29}{20} \div \frac{3}{20} = 9\frac{2}{3}$	c. $5 \div \frac{1}{3} = 15$	d. $7 \div 1\frac{2}{5} = 5$

3. She can fit 24 stickers on the cover.
Along the longer side of the notebook, she can fit (8 1/2 in) ÷ (1 1/4 in) = (17/2) ÷ (5/4) = (17/2) × (4/5) = 17 × 2/5 = 34/5 = 6 4/5 stickers. But, obviously she would not want to use parts of a sticker, so that means she can fit 6 stickers that way. Similarly, along the shorter side, we divide to find how many she can fit: (5 1/2 in) ÷ (1 1/4 in) = (11/2) ÷ (5/4) = (11/2) × (4/5) = 11 × 2/5 = 22/5 = 4 2/5 stickers. So, she can fit 6 stickers one way and 4 the other, a total of 24 stickers.

4.

Expression	the terms in it	coefficient(s)	Constants
$2x + 3y$	$2x$ and $3y$	2 and 3	none
$0.9s$	$0.9s$	0.9	none
$2a^4c^5 + 6$	$2a^4c^5$ and 6	2	6
$\frac{1}{6}f$	$\frac{1}{6}f$	$\frac{1}{6}$	none

5.

Serves (people)	6	12	18	24	30
butter	1/4 cup	1/2 cup	3/4 cup	1 cup	1 1/4 cups
sugar	1/2 cup	1 cup	1 1/2 cups	2 cups	2 1/2 cups
eggs	1	2	3	4	5
flour	3/4 cup	1 1/2 cups	2 1/4 cups	3 cups	3 3/4 cups

6. To serve 100 people you need 4 1/6 cups of butter, 8 1/3 cups of sugar, 17 eggs (it is rather hard to divide an egg), and 12 1/2 cups of flour. You can for example multiply by 4 the ingredient list for 24 people, and then take 1/3 of the ingredients for 12 people.

7.

a. $900 - \frac{1}{6} \cdot 72 = 888$	b. $23 + 3^4 = 104$	c. $\frac{100^3}{100^2} = 100$

8. We can see from the model that the difference of 9 years corresponds to 3 blocks, so one block is 3 years.
a. Marie is 12 years old. Tom is 21 years old.
b. The ratio of Marie's age to Tom's age is 4:7

Tom
9 years
Marie

9.

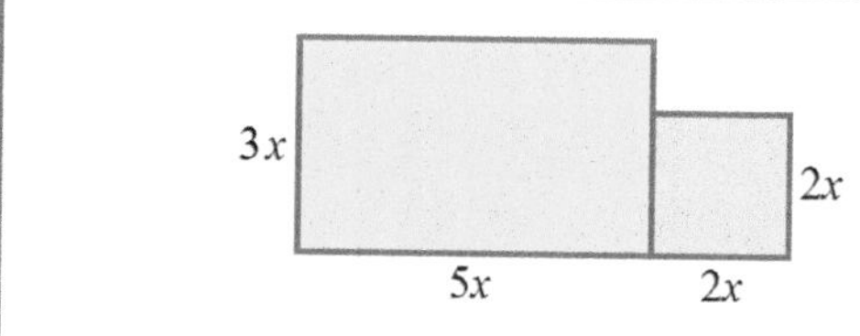

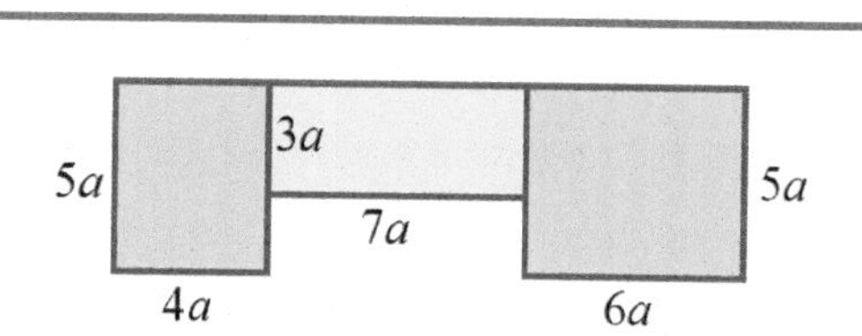

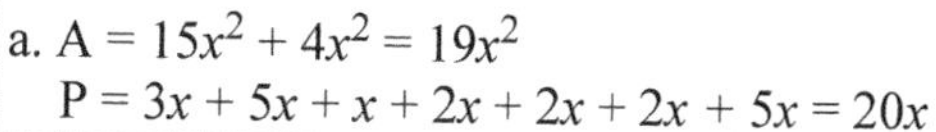

a. $A = 15x^2 + 4x^2 = 19x^2$
$P = 3x + 5x + x + 2x + 2x + 2x + 5x = 20x$

b. $A = 20a^2 + 21a^2 + 30a^2 = 71a^2$
$P = 4a + 5a + 4a + 7a + 6a + 5a + 6a + 2a + 7a + 2a = 48a$

Cumulative Review: Chapters 1 - 9

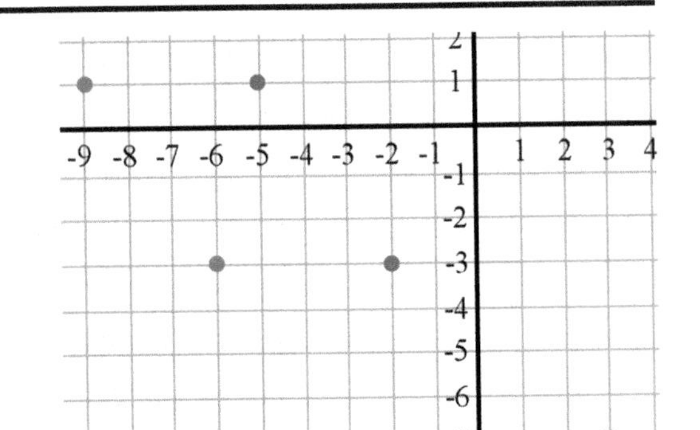

1. $(-5, 1) \rightarrow (-9, 1)$
   $(-2, -3) \rightarrow (-6, -3)$
   $(3, -7) \rightarrow (-1, -7)$

2. $\$180 \div 9 = \$20$; $4 \cdot \$20 = \$80$; $5 \cdot \$20 = \$100$
   Sam got \$80 and Matt got \$100.

3. a. 14
   b. 12

4. Answers will vary. Check students' answers.
   For example, 45, 90, 135, 180, and 225.

5.

a. $4 \times 0.0003 = 0.0012$	b. $0.2 \times 0.3 = 0.06$	c. $0.03 \times 1{,}000 = 30$

6.

a. $0.5x = 30$ $x = 60$	b. $0.01x = 2$ $x = 200$	c. $c + 1.1097 = 3.29$ $c = 2.1803$

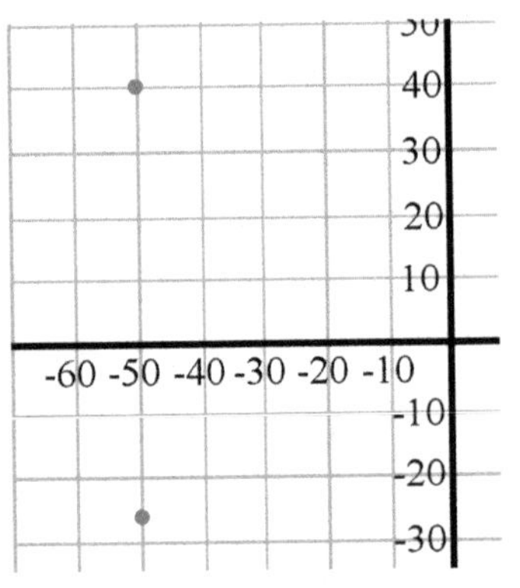

7. a. The game pieces are 65 units apart.
   b. Samantha said, "You missed by **60** units!"
   c. She has 2/6 = 33% of her pieces left.

8. The first flash drive costs $0.85 \times \$18 = \$15.30$ after the discount, and the second costs $\$20 \div 5 \times 4 = \$16$. So, the first one is the better deal.

9. Alice has 27 oranges and Michael has 9.
   After giving some to Beatrice, Alice has $90 \div 5 \times 2 = 36$ oranges.
   She gave 1/4 of those, or 9 oranges to Michael.

10. 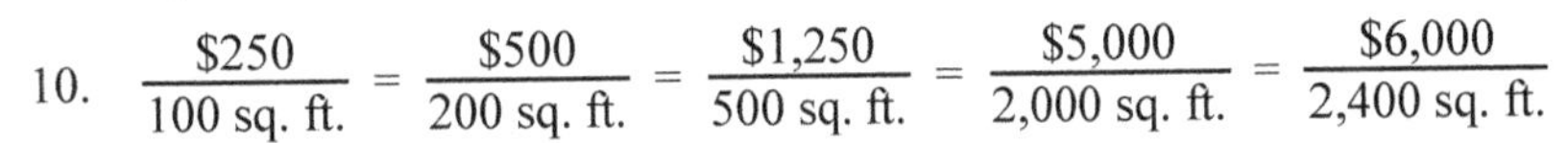

$$\frac{\$250}{100 \text{ sq. ft.}} = \frac{\$500}{200 \text{ sq. ft.}} = \frac{\$1{,}250}{500 \text{ sq. ft.}} = \frac{\$5{,}000}{2{,}000 \text{ sq. ft.}} = \frac{\$6{,}000}{2{,}400 \text{ sq. ft.}}$$

11.

a. $-2 + (-11) = -13$ $(-11) + 2 = -9$	b. $-1 + (-7) = -8$ $1 - 7 = -6$	c. $10 - 17 = -7$ $-10 - 17 = -27$	d. $7 - (-3) = 10$ $-3 - (-7) = 4$

12.

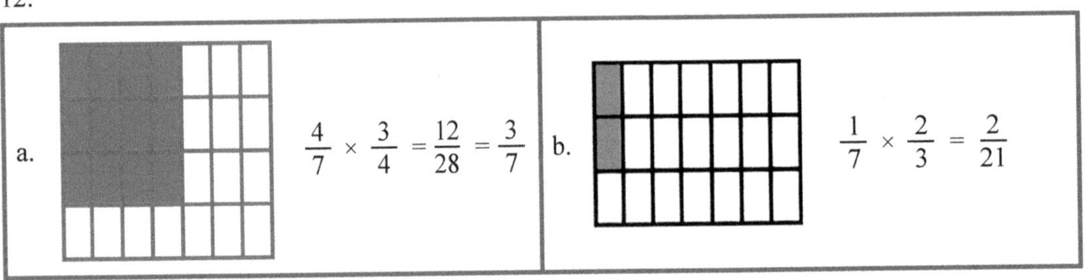

a. $\frac{4}{7} \times \frac{3}{4} = \frac{12}{28} = \frac{3}{7}$	b. $\frac{1}{7} \times \frac{2}{3} = \frac{2}{21}$

Cumulative Review: Chapters 1 - 10

1. Perimeter = 13.7 cm + 22.38 cm + 17.2 cm = 53.28 cm
   Area = 22.38 cm × 0.97 cm ÷ 2 = 10.8543 cm^2

2. a. It is 3,696 feet from Ben's home to his workplace (0.7 × 5,280 ft = 3,696 ft).
   b. Ben walks to work 4 days a week.
   Ben walks 4 × 48 × 3,696ft = 709,632 ft, which is 709,632 ft ÷ 5,280 ft = 134.4 miles.

3.

	m	dm	cm	mm
a. 7.82 m	7.82	78.2	782	7,820
b. 109 mm	0.109	1.09	10.9	109 mm

4. a. \$17.54 ÷ 3 = \$5.847
   b. 2.4 ÷ 0.05 = 48

5. a. The volume of the shoe box is 5,400 cm^3
   b.

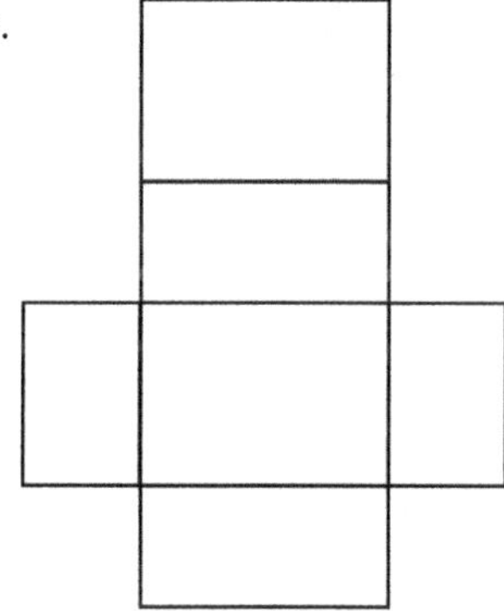

   c. Surface area = 2(18 × 25) + 2(12 × 25) + 2(12 × 18) = 1,932 cm^2

6. a. $x^5 - 5$
   b. $(2 - x)^3$
   c. $2(10 + y)$
   d. $\frac{s-2}{s^2}$

7. 32 − 5 = 27

8. a. $56x + 14 = 14(4x + 1)$
   b. $18u + 60 = 6(3u + 10)$

9.

a.	b.	c.
$y \div 50 = 60 \cdot 2$	$3x - x = 3 + 7$	$7x = 50$
$y \div 50 = 120$	$2x = 10$	$x = 50 \div 7$
$y = 120 \cdot 50$	$x = 10 \div 2$	$x = 7\ 1/7$
$y = 6{,}000$	$x = 5$	

10. {19, 21, 23}

11. 

12.

a. $\frac{5}{4} = 1.25$	b. $\frac{6}{7} = 0.857$	c. $\frac{19}{16} = 1.188$

13. 548.386 cm^2
In centimeters, the puzzle measures $8.5 \cdot 2.54 = 21.59$ cm by $10 \cdot 2.54 = 25.4$ cm.
Its area is $21.59 \cdot 25.4 = 548.386$ cm^2.

14. The oats cost \$2.53. This is easiest to calculate by changing 2 3/4 lb into the decimal 2.75 lb: $\$0.92 \cdot 2.75 = \$2.53$.

15.

Stem	Leaf
14	0 5 8
15	0 2 5 8 9
16	0 2 2 3 3
17	2 5
18	
19	0

CPSIA information can be obtained
at www.ICGtesting.com
Printed in the USA
LVOW06s1943190817
545615LV00005B/13/P

9 781942 715238